OZONE
Diplomacy

Published in cooperation with
World Wildlife Fund &
The Conservation Foundation

and

Institute for the
Study of Diplomacy,
Georgetown University

Policy Responses to Stratospheric
Ozone depletion
Haas, 1991

OZONE Diplomacy

New Directions
in Safeguarding
the Planet

Richard Elliot Benedick

Harvard University Press
Cambridge, Massachusetts, and London, England
1991

Printed in the United States of America
10 9 8 7 6 5 4 3 2 1

This book is printed on recycled and acid-free paper, and its binding materials
have been chosen for strength and durability.

Library of Congress Cataloging in Publication Data

Benedick, Richard Elliot.
Ozone diplomacy : new directions in safeguarding the planet /
Richard Elliot Benedick.
p. cm.
Published in cooperation with World Wildlife Fund & The
Conservation Foundation and Institute for the Study of
Diplomacy, Georgetown University.
Includes bibliographical references and index.
ISBN 0-674-65000-X (cloth). —ISBN 0-674-65001-8 (pbk.)
1. Vienna Convention for the Protection of the Ozone Layer (1985).
Protocols, etc., 1987. Sept. 15. 2. Ozone layer depletion—Law and
legislation. 3. Air—Pollution—Law and legislation. 4. Air
quality management—International cooperation. I. World Wildlife
Fund. II. Conservation Foundation. III. Georgetown University.
Institute for the Study of Diplomacy. IV. Title.
K3593.B46 1991
344'.046342—dc20
[342.4465342]
90-20879
CIP

To Helen

with very special thanks

Contents

Illustrations and Tables

Foreword

The negotiations leading to the Montreal Protocol on protection of the ozone layer, so perceptively described and analyzed by Ambassador Richard Benedick in this volume, are a major manifestation of a new dimension in diplomacy.

Throughout most of the twentieth century, diplomats have concentrated on questions of political and economic relations among nation-states, the traditional subjects of diplomacy. In the period following World War II other issues arose, spurred by the information revolution, the development needs of newly independent nations, and technical advances in nuclear energy and electronics. As the century closes, a third set of international problems—those relating to the health of the planet—is coming to the fore, presenting new challenges to diplomacy.

These problems will test the ability of governments and their diplomats to organize themselves for new dimensions in foreign relations, and to negotiate agreements that require departures from the traditional nation-state orientations of diplomacy toward patterns of global management still to be developed. The threatened depletion of the Earth's ozone layer is a prime example of such challenges. Similar items on the new agenda include global warming, the destruction of forests, the unprecedented extinction of plant and animal species, expanding desertification and soil erosion, and the pollution of such common resources as the oceans, the atmosphere, and the Antarctic continent.

Diplomats and governments alike may find this set of issues as difficult to manage and negotiate as any issues of peace or war. Because solutions to the problems of the environment must be global, they will present an unprecedented challenge to concepts of national sovereignty. As the ozone negotiations demonstrated, solutions will have to involve broad cooperation rather than competition among nation-states—perhaps eventually even new organizations that supersede the nation-state. Diplomatic efforts must also address divisive differences in perspective between the northern industrialized nations and the poorer nations of the Southern Hemisphere.

Environmental negotiations affect, too, the delicate relationship between the public and private sectors. Ways must be found to improve the coordination of the actions and policies of multinational corporations with those of governments. And the technical complexity of the problems makes them politically exploitable. These factors increase the difficulties of gaining public and legislative support for the kinds of hard decisions that must be made. The role of science advisers to governmental leaders becomes increasingly significant.

The tasks now facing diplomats and decision makers seeking to resolve world problems are complex but not hopeless. Five years ago, for example, few would have predicted the improvement in U.S.-Soviet relations. Patient and persistent diplomacy, multilateral pressures, and recognition of the futility of continuing conflict have brought results. Now diplomatic approaches to global environmental issues are beginning to yield results, even though the seriousness of the problems has been widely recognized only in the last decade. Such results become possible when a broad public recognition of the unacceptability of conditions creates pressure on governments to act. Only when that pressure makes itself felt in executive offices and legislatures will the new items on the global agenda assume equivalent prominence with matters of war and peace and the dramatic changes brought about by the communications revolution.

In this context, the achievement of the ozone treaty was a major breakthrough. In this age of high-technology communications and computers, it is easy to overlook the function of the diplomat. The story of the Montreal Protocol, however, demonstrates convincingly that negotiating success is still highly dependent on the imagination and skills of professional diplomats.

Richard Benedick has had years of diplomatic and negotiating experience as a career officer of the U.S. Foreign Service. As deputy assistant secretary of state for environment, health, and natural resource issues, he was assigned by Secretary of State George Shultz to coordinate U.S. preparations for a protocol on protecting the ozone layer and to be its chief U.S. negotiator. Designing and implementing an innovative negotiating strategy against heavy odds, Ambassador Benedick played a crucial leadership role in the international process. In recognition of his work in securing this historic treaty, he received in 1988 the highest career public service honor, the Presidential Distinguished Service Award. Richard Benedick's contribution is an example of how one individual's vision and commitment can be critical to effective modern diplomacy.

David D. Newsom
Former U.S. Under Secretary of State for Political Affairs

Preface

In January 1985 I led a small U.S. delegation to a little-noticed meeting in Geneva, where we failed to achieve an international agreement. The event attracted only perfunctory attention in the press, and its unremarkable results occasioned no diplomatic ripples in national capitals, including Washington.

The Geneva meeting was a session of the cumbersomely labeled Ad Hoc Working Group of Legal and Technical Experts for the Preparation of a Global Framework Convention for the Protection of the Ozone Layer. This collection of diplomats, environmental officials, and government lawyers from some two dozen countries had struggled with its task for three years under the auspices of a small UN agency, the United Nations Environment Programme. Its objective was to craft an international accord based on an unproved scientific theory that certain anthropogenic chemicals could destroy a remote gas in the stratosphere and thereby possibly bring harm to human health and the environment in the distant future.

The negotiators failed to reach consensus on how to regulate these nearly ubiquitous chemicals. They did, however, agree on the need for a treaty providing for international cooperation in research, monitoring, and exchange of information on factors affecting stratospheric ozone. More important, the Vienna Convention on Protection of the Ozone Layer, which was signed in March 1985, also established a procedure for reopening negotiations on a protocol that could mandate future international controls.

Only 30 months later, in September 1987, the signing of the Montreal Protocol on Substances That Deplete the Ozone Layer made headlines around the world. And by 1989 protection of the ozone layer was figuring prominently in discussions among the world's political leaders. Chlorofluorocarbons (CFCs) and ozone had become, literally, household words. What happened between the publication of scientific theories in 1974 and the signing of a landmark treaty in 1987, together with the subsequent issues that came up in implementing that treaty, is both fascinating and instructive.

When, in the summer of 1986, I was asked to continue my work on ozone by leading the U.S. preparations for international negotiations on an ozone protocol, the task seemed formidable. Most observers in and out of government believed at that time that an agreement on international regulation of CFCs would be impossible to reach. The issues were staggeringly complex, involving interconnected scientific, economic, technological, and political variables. The science was still speculative, resting on projections from evolving computer models of imperfectly understood stratospheric processes—models that yielded varying, sometimes contradictory, predictions of potential future ozone losses each time they were further refined. Moreover, existing measurements of the ozone layer showed no depletion, nor was there any evidence of the postulated harmful effects.

Powerful forces were arrayed in opposition to any treaty. U.S. and foreign industrialists were strongly hostile to regulation, claiming that the economic and social costs would be unacceptable and that technical solutions were out of reach; many corporate officials scoffed at the idea of basing costly controls on unproved theories. These views were echoed by the 12-member European Community, Japan, and the Soviet Union, which together accounted for nearly two-thirds of the world's production of ozone-depleting chemicals and whose cooperation was therefore essential for any treaty to be effective. Negotiation with the European Community was further complicated by its emerging political and economic union; it was often unclear whether the European Commission in Brussels or the sovereign member states would have responsibility for implementing critical decisions relevant to the protocol. In addition, most developing countries, whose large and rapidly growing populations represented enormous potential demand for CFCs, showed little interest in participating in negotiations aimed at curtailing products that seemed almost synonymous with the standards of living to which they aspired.

Managing the negotiations also involved working closely with mutually antagonistic environmental advocacy groups and industry, as well as with several congressional committees and the increasingly interested international media. The activities of industry and environmental groups strongly influenced the course of events, and their representatives actually attended intergovernmental negotiations. The U.S. Congress held many hearings on ozone policy and on the progress of the negotiations, congressional staff attended the international meetings, and Congress passed resolutions and introduced legislation that affected the negotiations. And just as the American negotiators were beginning to make substantial progress on the diplomatic front, a domestic antiregulatory backlash within the U.S. administration almost succeeded in overturning the U.S. position.

In the chapters that follow I analyze the policymaking and diplomatic process, from the first scientific hypotheses in the 1970s, through negotiation of the 1985 Vienna Convention and the 1987 Montreal Protocol, to the crucial implementation issues that arose after the protocol's signing. This book thus attempts to provide a comprehensive account of how the international community approached an unprecedented global ecological threat, one that required governments to balance distant but possibly catastrophic dangers against the very real short-run economic dislocations that would be caused by preventive measures. Such situations will probably become the norm as greenhouse warming, deforestation, toxic substances, air and water pollution, mass extinction of species, and similar issues move up on the foreign policy agenda.

The ozone experience offers lessons for policymakers and diplomats in confronting this new generation of global environmental challenges. As was the case with ozone, entrenched economic interests, scientific uncertainties, technological limitations, and political timidity will continue to be formidable obstacles to action by governments and industry. In implementing foreign policy, diplomats will have to work increasingly closely, as we did, with scientists, environmentalists, citizens' groups, and industrialists. Multilateral diplomacy will assume much greater importance. And new forms of international cooperation and aid will be needed to enlist developing nations in common efforts to protect the global environment while promoting sustainable economic growth. In effect, the Montreal Protocol process can be examined as a paradigm for new diplomatic approaches to new kinds of international challenges.

The history of the ozone issue in general—and of the Montreal Protocol in particular—is also a case study of the diplomatic craft. The formal records of international meetings do not tell the whole story of what ensued. They cannot convey either the atmosphere of the negotiations or the dynamic and informal interchange among individuals and ideas. The following pages, though by no means a personal memoir, attempt to provide at least some of this flavor, including a few vignettes of the often-unrecorded serendipitous events (a proposal at lunch, an informal exchange, an overheard remark) that can change the course of a negotiation outside the formal plenary sessions. Professional diplomats and students of negotiations may derive insights from the story of how obstacles were overcome in achieving an "impossible" accord. Many of the key provisions of the protocol, described in detail, represent innovative solutions to complicated equity and technical issues that have relevance for future negotiations.

The following account inevitably represents a primarily American perspective. However, I have drawn extensively not only on personal notes of numerous informal conversations with non-U.S. participants and observers but also on foreign documentation; both published sources and internal government memoranda supplement official reports of the United Nations and several governments. Scientific, environmental, and industry journals and publications were another rich source of data and ideas. Finally, the press provided an illuminating commentary on the process, and I refer to its coverage often as supplementary documentation of events or occurrences that were otherwise attributable only to personal observations or conversations.

After the signing of the Montreal Protocol, I was fortunate to be assigned by the Department of State as a senior fellow of World Wildlife Fund & The Conservation Foundation. Although the State Department regularly sends career diplomats to think-tanks and to university faculties of political science, economics, or government, this was the first such appointment to a private environmental organization—perhaps a symbol of the coming of age of environment in our foreign policy? During this period I frequently spoke in public on the ozone experience and participated in numerous symposia, roundtables, and conferences in the United States and abroad. These occasions afforded an opportunity to test ideas and interpretations of the ozone history on knowledgeable observers from universities, the scientific community, industry, environmental groups, and governments and international organizations. Many insights derived from these exchanges are reflected in the following pages.

I am grateful to the Department of State for affording this opportunity to reflect on the ozone negotiations, thereby deepening my own understanding of a process that, at the time, rushed by with breakneck speed. World Wildlife Fund & The Conservation Foundation, under the exceptional leadership of Kathryn S. Fuller and her predecessor William K. Reilly (now administrator of the U.S. Environmental Protection Agency), provided a congenial home base and generous logistical support. Most of the basic writing was accomplished in the splendid setting of the National Center for Atmospheric Research, which, thanks to the invitation of Michael Glantz, offered a haven from the distractions of Washington.

I am indebted to many people for their time, interest, encouragement, and ideas. The late Walter Orr Roberts, president emeritus of the University Corporation for Atmospheric Research, was a very special source of inspiration. Perceptive and critical reviewers of the manuscript in various phases have included Guy Brasseur, Ralph Cicerone, John Firor, Michael Glantz, and Stephen Schneider at the National Center for Atmospheric Research; Daniel Albritton and Susan Solomon at the National Oceanic and Atmospheric Administration's Aeronomy Laboratory in Boulder, Colorado; J. Clarence Davies, Robert McCoy, Julia Moore, Jack Noble, and Bradley Rymph at World Wildlife Fund & The Conservation Foundation; Richard Scribner and Margery Boichel Thompson at Georgetown University; Kathleen Courrier and Jessica Tuchman Mathews at the World Resources Institute; Alan Miller at the Center for Global Change, University of Maryland; Christopher Joyner at George Washington University; and Lincoln Bloomfield at the Massachusetts Institute of Technology. Sincere appreciation also goes to Mary Downton, Fannie Mae Keller, Maria Krenz, Janice Marrow, and Jan Stewart for their valued logistical support.

Finally, my special thanks to Ann Hawthorne of Harvard University Press for her sensitive and enthusiastic counsel in improving the manuscript.

With all of this help, the responsibility for any lapses and errors is mine alone; the views and interpretations of events are also personal, and this book does not by any means represent an official U.S. government history.

The achievement of the Vienna Convention and the Montreal Protocol involved the talents and efforts of many individuals. I feel particularly grateful to the members of U.S. delegations to the negotiations and to those who sustained us in our work. To Secretary of State George Shultz, Deputy Secretary of State John Whitehead, Assistant Secretaries of State John Negroponte and James Malone, and Environmental Protection Agency Administrator Lee Thomas, my appreciation for their unwavering support and leadership. And to the members of U.S. delegations and special missions at various stages in the ozone history, my heartfelt thanks and congratulations on a job well done: from the State Department—Suzanne Butcher, Robert Coe, Scott Hajost, Deborah Kennedy, John Rouse, Andrew Sens, Richard J. Smith, James Timbie; from the Environmental Protection Agency—Stephen Anderson, Eileen Claussen, John Hoffman, William Long, James Losey, Stephen Seidel, Stephen Weil,

Dwain Winters; from the National Aeronautics and Space Administration—Richard Stolarski, Robert Watson; from the National Oceanic and Atmospheric Administration—Daniel Albritton, J. R. Spradley; from the Office of the U.S. Trade Representative—Robert Reinstein; from the President's Council of Economic Advisers—Stephen DeCanio; from the Department of Energy—Edward Williams; from the Department of Commerce—Edward Shykind; from the Department of Justice—Thomas Hookano; from the Department of the Interior—Martin Smith. We shared moments of discouragement and triumph.

. . . this most excellent canopy, the air, look you,
this brave o'erhanging firmament, this majestical
roof fretted with golden fire, why, it appeareth
nothing to me but a foul and pestilent congregation
of vapors.

Shakespeare, *Hamlet,* Act II, scene 2

Abbreviations

CFC	chlorofluorocarbon
CT	carbon tetrachloride
EC	European Community
EC Commission	Commission of the European Community
EPA	U.S. Environmental Protection Agency
HCFC	hydrochlorofluorocarbon
HFC	hydrofluorocarbon
MC	methyl chloroform
UNEP	United Nations Environment Programme
WMO	World Meteorological Organization

Lessons from History

1

On September 16, 1987, a treaty was signed that was unique in the annals of international diplomacy. Knowledgeable observers had long believed that this particular agreement would be impossible to achieve because the issues were so complex and arcane and the initial positions of the negotiating parties so widely divergent. Those present at the signing shared a sense that this was not just the conclusion of another important negotiation, but rather a historic occasion. It was hailed as "the most significant international environmental agreement in history," "a monumental achievement," and "unparalleled as a global effort."[1]

The Montreal Protocol on Substances That Deplete the Ozone Layer mandated significant reductions in the use of several extremely useful chemicals. At the time of the treaty's negotiation, chlorofluorocarbons (CFCs) and halons were rapidly proliferating compounds with wide applications in thousands of products, including refrigeration, air conditioning, aerosol sprays, solvents, transportation, plastics, insulation, pharmaceuticals, computers, electronics, and fire fighting. Scientists suspected, however, that as these substances were released into the atmosphere and diffused to its upper reaches, they might cause future damage to a remote gas—the stratospheric ozone layer—that shields life on Earth from potentially disastrous levels of ultraviolet radiation.

By their action, the signatory countries sounded the death knell for an important part of the international chemical industry, with implications for billions of dollars in investment and hundreds of thousands of jobs in related sectors.[2] The protocol did not simply prescribe limits on these chemicals based on "best available technology," which had been a traditional way of reconciling environmental goals with economic interests. Rather, the negotiators established target dates for replacing products that had become synonymous with modern standards of living, even though the requisite technologies did not yet exist.

"Politics," Lord Kennet stated in special hearings on the accord in the British House of Lords a year later, "is the art of taking good decisions on insufficient evidence."[3] Perhaps the most extraordinary aspect of the treaty was its imposition of substantial short-term economic costs to protect human health and the environment against unproved future dangers—dangers that rested on scientific theories rather than on firm data.

At the time of the negotiations and signing, no measurable evidence of damage existed. Thus, unlike environmental agreements of the past, the treaty was not a response to harmful developments or events but rather a preventive action on a global scale. Indeed, the Montreal Protocol added a new dimension to the 1972 Declaration of the United Nations Conference on the Human Environment at Stockholm, which appealed to nations "to ensure that activities within their jurisdiction of control do not cause damage to the environment of other States"—a responsibility that had been more often an ideal than a reality.[4]

As the international community and national political leaders grapple with other environmental problems involving scientific uncertainty, long-term risks, and multilateral cooperation, the model of ozone diplomacy offers a valuable precedent. In an effort to draw lessons from this experience, the following pages analyze the scientific, economic, and political background to the Montreal Protocol, the process involved in achieving international consensus, and the developments following the treaty's signing that were crucial to its successful implementation.

Breaking New Ground

The Montreal Protocol was a response to a new type of problem facing the modern world. In the past two decades there has been a growing realization that ecological dangers can imperil the security of all peoples. Ozone layer depletion, climate change, destruction of tropical rain forests, toxic wastes, pollution of oceans and fresh water, massive loss of biological diversity—these issues are moving to the top of the world's agendas, and they will increasingly dominate international relations in the 1990s. The drafting of the Montreal Protocol was only the first of a series of international events in the late 1980s that brought together national leaders to consider these global threats. Symbolic of the awakening political awareness was the unprecedented concern over the deteriorating environment expressed at the 1989 and 1990 economic summit meetings of the major industrial democracies. Twenty years after the Stockholm conference, world leaders will convene in June 1992 in Brazil at the United Nations Conference on Environment and Development, which will generate crucial guidelines for international approaches to these problems over the coming decades.

This new generation of ecological issues exemplifies the interconnectedness of life and its natural support systems on Earth. Modern scientific discoveries are revealing that localized activities can have global consequences and that dangers can be slow and perhaps barely perceptible in their development, yet with long-term and virtually irreversible effects. The concept is not obvious: a perfume spray in Paris helps to destroy an invisible gas in the stratosphere and thereby contributes to skin cancer deaths and species extinction half a world away and several generations in the future.

Neither traditional environmental law nor traditional diplomacy offers guidelines for confronting such situations. Environmental problems of the past were normally localized or regional, and their effects were self-evident. Such events could be addressed largely through unilateral actions, national legislation, and occasional international treaties, all based on unmistakable evidence of damage.[5]

However, if the international community is to respond effectively to the new environmental challenges, governments must undertake coordinated actions while some major questions remain unresolved—and before damage becomes tangible and thereby possibly irremediable. The negotiators of the Montreal Protocol dealt with dangers that could touch every nation and all life on Earth, over periods far beyond the normal time horizons of politicians. But the potentially grave consequences could be neither measured nor predicted with any certainty at the time.

The Montreal Protocol thus became a prototype for an evolving new form of international cooperation. In achieving the treaty, consensus was forged and decisions were made on a balancing of probabilities. And the risks of waiting for more complete evidence were finally deemed to be too great.

As U.S. delegation head, I summed up the case for action in a plenary address at one of the early negotiating rounds:

> When we build a bridge, we build it to withstand much stronger pressures than it is ever likely to confront. And yet, when it comes to protecting the global atmosphere, where the stakes are so much higher, the attitude [of some people] seems to be equivalent to demanding certainty that the bridge will collapse as a justification for strengthening it. If we are to err in designing measures to protect the ozone layer, then let us, conscious of our responsibility to future generations, err on the side of caution.[6]

Challenges and Leadership

Ironically, as time has passed, and as disturbing new scientific discoveries have vindicated the concerns of the protocol's designers, the events at Montreal have acquired an aura of inevitability. It all seems easy in retrospect. But memories are short.

Both before and during the negotiations, major uncertainties surrounded such fundamental questions as the possible degree of future damage to stratospheric ozone, the extent to which CFCs and other chemicals were responsible, the prospective growth of demand for these chemicals, the significance of any adverse effects from ozone layer depletion, and the length of time before serious harm might occur. Many governments whose cooperation was needed displayed attitudes ranging from indifference to outright hostility. International industry was strongly opposed to regulatory action. And the gaps separating the negotiating parties often seemed unbridgeable.

The problem of protecting the stratospheric ozone layer presented an unusual challenge to diplomacy. Military strength was irrelevant to the situation. Economic power also was not decisive. Neither great wealth nor sophisticated technology was necessary to produce large quantities of ozone-destroying chemicals. Traditional notions of national sovereignty became questionable when local decisions and activities could affect the well-being of the entire planet. Because of the nature of ozone depletion, no single country or group of countries, however powerful, could effectively solve the problem. Without far-ranging cooperation, the efforts of some nations to protect the ozone layer would be undermined.

In this unique and complex situation, a small and previously little-known UN agency, the United Nations Environment Programme (UNEP), together with the U.S. government, assumed primary leadership roles.

They faced an uphill battle all the way. How the consensus for international action was achieved is a case study in modern diplomacy.

Elements of Success

The success at Montreal can be attributed to no single prime cause. Rather, a combination of planning and chance, of key factors and events, made an agreement possible. Nevertheless, certain elements, discussed in more detail in the following chapters, are particularly relevant for policymakers struggling to find viable approaches to such comparable issues as climate change.

First and foremost was the indispensable *role of science* in the ozone negotiations. Scientific theories and discoveries alone, however, were not sufficient to influence policy. The best scientists and the most advanced technological resources had to be brought together in a cooperative effort to build an international scientific consensus. Close collaboration between scientists and government officials was also crucial. Scientists were drawn out of their laboratories and into the negotiating process, and they had to assume an unaccustomed and occasionally uncomfortable shared responsibility for the policy implications of their findings. For their part, political and economic decision makers needed to understand the scientists, to fund the necessary research, and to be prepared to undertake internationally coordinated actions based on realistic and responsible assessments of risk.

Second, the *power of knowledge and of public opinion* was a formidable factor in the achievement at Montreal. A well-informed public was the prerequisite to mobilizing the political will of governments and to weakening industry's resolve to defend the chemicals. The findings of scientists had to be made accessible and disseminated. Legislative hearings helped in airing scientific opinion and policy alternatives. The media, particularly press and television, played a vital role in bringing the issue before the public and thereby stimulating political interest. Both UNEP and the U.S. government undertook major public education campaigns on ozone and CFCs, using traditional diplomatic channels as well as various communications media. These efforts, aimed at both governments and citizenries, helped influence several countries to change their initial positions on the need for regulations.

Third, because of the global scope of the issues, the *activities of a multilateral institution* were critical to the success of the negotiations. UNEP commanded respect for its commitment and its sensitivity to national interests, particularly in the developing countries. UNEP was indispensable in mobilizing data and informing world public opinion, as well as during the negotiating and implementing phases. It was UNEP—inviting, cajoling, and pressuring governments to the bargaining table—that broadened the protocol to a global dimension. UNEP also provided an objective international forum, free of the time-consuming debates on irrelevant political issues that have often marred the work of other UN bodies. The strong personality of its executive director, Mostafa Tolba, an Egyptian scientist, was a driving force in achieving the eventual consensus. Rather than merely playing a mediating role between opposing sides, Tolba risked taking personal positions, advancing views and concerns that might otherwise have been overlooked and making UNEP in a sense the advocate for governments and populations not present at the negotiations. In sum, UNEP went far beyond a traditional secretariat function: it was a model for effective multilateral action.

Fourth, an *individual nation's policies and leadership* made a major difference. The United States undertook such leadership in achieving international agreement on ozone protection. The U.S. government set the example by being the first to take regulatory action against the suspect chemicals. Later, it developed a comprehensive global plan for protecting the ozone layer and tenaciously campaigned for its international acceptance through bilateral and multilateral initiatives. The staff of the U.S. Environmental Protection Agency (EPA) labored tirelessly to develop volumes of analyses on all aspects of the problem that gradually had an impact on skeptics both at home and abroad. A negotiating strategy devised by the U.S. Department of State not only employed customary diplomacy but also capitalized on the expertise of EPA and the National Aeronautics and Space Administration (NASA), as well as involving Congress, environmental groups, industry, and the media.

The progressive U.S. policies on ozone, reinforced by the size and importance of the United States as the largest single producer and consumer of CFCs and halons, enabled it to play a formative central role in the drive for international controls. Other nations, however, also had a share of leadership and influence. The Federal Republic of Germany, for example, was instrumental in turning around the European Community's initially negative position. And the ozone history demonstrated that even countries with small or negligible shares of production or consumption could exert a disproportionate leverage on the course of events: Canada and Norway were consistent leaders from the very start in pushing for strong measures to protect the ozone layer, and Australia, Finland, and New Zealand played major roles in decisions to strengthen the treaty in 1990.

Fifth, *private-sector organizations*—industry and citizens' groups—participated actively, and usually with opposing objectives, in the ozone diplomacy. Environmental organizations warned the general public of the risks, pressured governments to act, and promoted research and legislation. Industrial associations and individual firms also mounted efforts to influence public and official opinion; industry's outlook and actions were crucial, since the technical solutions to the problems ultimately depended on its cooperation. Representatives of industrial and environmental groups attended the ozone negotiations as observers, offering their differing perspectives to both negotiators and media representatives.

Sixth, the *process* involved in reaching the ozone accord was itself a determining factor. Subdividing this complex problem into more manageable components during a prenegotiating phase proved invaluable. Well before the opening of formal negotiations for an ozone protocol, informal scientific and economic workshops generated creative ideas and laid the foundation for international consensus. This fact-finding process was instrumental in overcoming the arguments for delaying action. The extensive preliminary scientific and diplomatic groundwork enabled the subsequent negotiations to move forward relatively rapidly.

Finally, unlike traditional international treaties that seek to cement a status quo, the Montreal Protocol was deliberately designed as a *flexible and dynamic instrument.* Relying on periodic scientific, economic, environmental, and technological assessments, it can be readily adapted to evolving conditions. The treaty's provisions, as described in the chapters that follow, incorporate innovative solutions to equity and technical problems. Indeed, the protocol's essence is that, far from being a static solution, it constitutes an ongoing process. This factor was clearly demonstrated in the events leading to the substantial strengthening and modification of the treaty at the Second Meeting of Parties in London in June 1990. As UNEP's Tolba observed: "The mechanisms we design for the Protocol will—very likely—become the blueprint for the institutional apparatus designed to control greenhouse gases and adaptation to climate change."[7]

* * *

In conclusion, the ozone accord broke new ground in its treatment of long-term risks and in its reconciliation of difficult scientific, economic, and political issues. In view of the many complexities and uncertainties involved, the Montreal Protocol was achieved with astonishing rapidity. The signing occurred 13 years after the first scientific hypothesis on the ozone layer was published in 1974—and only nine months after formal diplomatic negotiations began in December 1986. The protocol entered into force as planned on January 1, 1989, and within the next 19 months a total of 62 countries had ratified it. (See Appendix D for a complete list of signing and ratifying countries as of August 2, 1990.)

What President Reagan characterized, when he signed the U.S. instrument of ratification in April 1988, as "an extraordinary process of scientific study . . . and of international diplomacy" may well prove to be a paradigm for a new form of global diplomacy.[8]

The Science: Models of Uncertainty

2

The Montreal Protocol was the result of research at the frontiers of science combined with a unique collaboration between scientists and policymakers. Unlike any previous diplomatic endeavor, it was based on continually evolving theories, on state-of-the-art computer models simulating the results of intricate chemical and physical reactions for decades into the future, and on satellite-, land-, and rocket-based monitoring of remote gases measured in parts per trillion. An international agreement of this nature could not, in fact, have occurred at any earlier point in history.

Disturbing Theories

The existence of ozone was unknown before 1839.[1] An unstable form of oxygen composed of three, rather than the customary two, atoms of oxygen, ozone has been characterized as "the single most important chemically active trace gas in the earth's atmosphere."[2] This significance derives from two singular properties. First, certain wavelengths of ultraviolet radiation are absorbed by the very thin "layer" of ozone molecules surrounding Earth, particularly in the upper part of the atmosphere known as the stratosphere, approximately 6 to 30 miles above the surface. If these biologically active ultraviolet (UV-B) lightwaves were to reach the planet's surface in excessive quantities, they could damage and cause mutations in human, animal, and plant cells.[3] Second, the distribution of ozone throughout different altitudes could influence the temperature structure and circulation patterns of the stratosphere and thus have major implications for climate around the world.[4] It is no exaggeration to conclude that the ozone layer, as currently constituted, is essential to life as it has evolved on Earth.

In 1973 two University of Michigan scientists, Richard Stolarski and Ralph Cicerone, were exploring the effects of possible chemical emissions from National Aeronautics and Space Administration rockets. Their research, published in 1974, indicated that chlorine released in the stratosphere could unleash a complicated chemical process that would continually destroy ozone for several decades. A single chlorine atom, through a catalytic chain reaction, could eliminate tens of thousands of the ozone molecules.[5] This theory, though interesting, did not at first seem alarming, because the potential release of chlorine from space rocketry would be inconsequential.

In 1974 Mario Molina and Sherwood Rowland at the University of California, Irvine, became intrigued with some peculiar properties of a family of widely used anthropogenic chemicals, the chlorofluorocarbons. Molina and Rowland discovered that, unlike most other gases, CFCs are not chemically broken down or rained out quickly in the lower atmosphere but rather, because of their exceptionally stable chemical structure, persist and migrate slowly up to the stratosphere. Depending on their individual structure, different CFCs can remain intact for many decades to several centuries. The two researchers concluded that CFCs are eventually broken down by solar radiation and in the process release large quantities of chlorine into the stratosphere.[6]

The combined implications of these two independently-arrived-at hypotheses were deeply disturbing. The researchers had not anticipated any link between CFCs and ozone depletion. There had been no prior suspicion that CFCs were harmful to the environment. Indeed, following their invention in the 1930s, CFCs had seemed an ideal chemical. They had been thoroughly tested by customary standards and found to be safe. The possibility that dangers could originate many miles above Earth's surface was never considered.

CFCs are unusually stable, nonflammable, nontoxic, and noncorrosive—qualities that make them extremely useful in many industries, where they often replaced other chemicals, such as ammonia in refrigerators, whose dangers were known. Because CFCs vaporize at low temperatures, they are energy-efficient coolants in refrigerators and air conditioners, as well as effective propellants in spray containers for cosmetics, household products, pharmaceuticals, and cleaners. They are also excellent insulators and are standard ingredients in the manufacture of a wide range of rigid and flexible plastic-foam materials. Their nonreactive properties make them seemingly perfect solvents for cleaning microchips and telecommunications equipment and for use in a myriad of other industrial applications. And, as an added bonus, CFCs are inexpensive to produce.

The 1974 theories came, therefore, as an economic as well as environmental bombshell. Because new uses had continually been found for CFCs, their production had soared from 150,000 metric tons in 1960 to over 800,000 metric tons by 1974.[7] Because of their long lifetimes, as much as nine-tenths of all CFCs ever emitted was still in the atmosphere. Millions of tons of prior-year CFC production were thus already en route, so to speak, to their fatal stratospheric rendezvous with ozone. Even if CFC emissions were to level off or decline, chlorine would continue to accumulate in the stratosphere for decades. If the chlorine-ozone breakdown process actually occurred in the stratosphere, as indicated by the laboratory results, some future depletion of the vitally important ozone layer seemed inescapable.

Initial Scientific Response

The U.S. scientific community reacted to the 1974 theories by mounting a major research campaign, involving the National Academy of Sciences and a growing number of prominent chemists, meteorologists, physicists, and space scientists from NASA, the National Oceanic and Atmospheric Administration (NOAA), and leading universities. The next several years were marked by intense professional and personal disputes within the scientific community.[8] Although a series of laboratory and modeling studies resulting from these activities confirmed the validity of the chlorine-ozone linkage, they could not prove definitively that it described what was actually going on in the stratosphere.[9]

The research complexities involved were enormous. Ozone amounts to considerably less than one part per million of the total atmosphere, with 90 percent of it concentrated above six miles in altitude. The intrinsically unstable ozone molecules are constantly being created and destroyed by complex natural forces involving solar radiation and interactions with even more minute quantities of other gases. Ozone concentrations fluctuate wildly as a result of natural causes on a daily, seasonal, and solar-cyclical basis; indeed, during the 1960s average ozone levels actually increased. Furthermore, great geographic variations in ozone abundance occur over different latitudes, as well as at different altitudes in the atmosphere. Amid all these large-scale fluxes, scientists thus faced an enormous challenge in attempting to detect a minuscule "signal" of the beginning of the postulated long-term downward trend in stratospheric ozone concentrations.[10]

The international chemical industry vigorously denied any connection between the condition of the ozone layer and increasing sales of CFCs. Industry forces quickly mobilized their own research and public relations efforts to cast doubt on the theory.[11] Industry's position was summarized in the 1974 congressional testimony of a top Du Pont executive: "The chlorine-ozone hypothesis is at this time purely speculative with no concrete evidence . . . to support it." He went on to say, in a pledge that would come back to haunt the industry 14 years later, "If creditable scientific data . . . show that any chlorofluorocarbons cannot be used without a threat to health, Du Pont will stop production of these compounds."[12]

As scientists' understanding increased, they began designing ever more complicated models to simulate the stratospheric interplay between radiative, chemical, and dynamic processes such as wind and temperature. Researchers discovered that other gases might also significantly affect stratospheric ozone. Growing concentrations of carbon dioxide and methane, for instance, could offset the projected impact of chlorine, while nitrogen compounds could influence the reaction in either direction.[13] Assumptions about the growth rates of these and other gases, as well as about future demand for CFCs, were obviously crucial to projections of ozone concentrations.

Methane originates in a wide range of sources, including the burning of biomass (as in clearing of forests and slash-and-burn farming), industrial activities (mining, oil and gas exploration), and bacterial processes occurring in swamps, rice paddies, and the digestion of livestock and termites. Hence the volume of methane is linked with agricultural practices, deforestation, and energy demand, which are themselves driven by global population growth, economic development, and social behavior.

Emissions of carbon dioxide (through fossil fuel combustion and deforestation) and CFCs are also influenced by industrial and energy policies, along with population-related demand. The increase of nitrogen compounds in the atmosphere is affected by such factors as agriculture (fertilizers), industry (steel mills), and transportation (automobiles and aircraft).

Thus, the ozone-depletion theory began by the late 1970s to assume much more complicated dimensions. Indeed, ozone protection became an exemplar of the new generation of environmental issues referred to in Chapter 1. To understand what was happening to the ozone layer, researchers needed to go far beyond atmospheric chemistry; they had to examine the planet as a system of interrelated physical, chemical, and biological processes taking place on land, in water, and in the atmosphere, processes that are themselves influenced by economic, political, and social forces.[14]

Not surprisingly, in the years following the initial hypotheses, the evolving understanding of the chemical processes and the interplay of so many elements produced wide fluctuations in the predicted effects of CFC emissions. Various model projections of global-average ozone depletion 50 to 100 years in the future began at about 15 percent in 1974, fell to around 8 percent in 1976, climbed again to almost 19 percent in 1979, and then dropped steadily to only about 3 percent by 1983 (see Figure 2.1).[15] These swings began to affect the credibility of the science and to

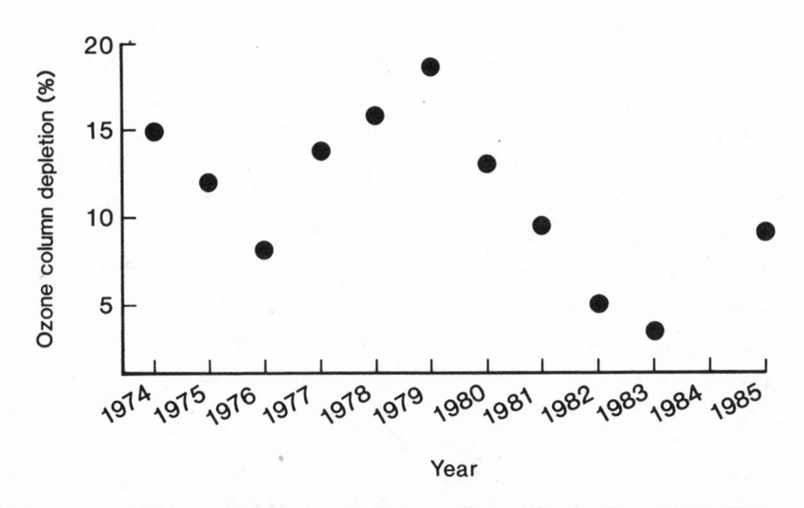

Figure 2.1 Various predictions of ozone layer depletion, 1974–1985.
Sources: Adapted from National Research Council, *Causes and Effects of Changes in Stratospheric Ozone: Update 1983* (Washington, D.C.: National Academy Press, 1984); World Meteorological Organization, *Atmospheric Ozone 1985* (Geneva, 1986).

dampen both public and official concern about the urgency of the problem.

A Landmark International Report

In late 1984, in a conscious effort "to provide governments around the world with the best scientific information currently available on whether human activities represent a substantial threat to the ozone layer," a remarkable cooperative international scientific venture was launched.[16] This integrative research was cosponsored by NASA, NOAA, the U.S. Federal Aviation Administration, the United Nations Environment Programme, the World Meteorological Organization (WMO), the West German Ministry for Research and Technology, and the Commission of the European Communities.

Coordinated by NASA, the work occupied approximately 150 scientists of various nations for over a year. The result, published by WMO and UNEP in 1986, was the most comprehensive study of the stratosphere ever undertaken: three volumes containing nearly 1,100 pages of text plus 86 reference pages listing hundreds of articles. The treatise covered the chemical, physical, and radiative processes affecting ozone; the natural and industrial sources of relevant chemicals; the existing measurements of the composition and temperature of the stratosphere; and model predictions of ozone and climate change based on different assumptions of future emissions of various trace gases. It balanced laboratory studies, field measurements, and theoretical models, and, unlike most previous reports, it analyzed changes and influences of the many relevant gases collectively rather than in isolation. The study also separated the effects of anthropogenic trace gases on the atmosphere from such naturally occurring phenomena as solar activity and volcanic eruptions.

A major finding of the WMO/UNEP report was that accumulations of CFCs 11 and 12 in the atmosphere had nearly doubled from 1975 through 1985.[17] Since actual production of these chemicals had stagnated over this period, these measurements confirmed the existence of a potential for large future increases in stratospheric concentrations of these long-lived substances, particularly if the growth rate of their emissions were to resume.

The WMO/UNEP assessment predicted that continued emissions of CFCs 11 and 12 at the 1980 rate could, through release of chlorine in the stratosphere, reduce the ozone layer by about 9 percent on a global average by the last half of the twenty-first century, with even greater seasonal and latitudinal declines.[18] As a result, higher levels of biologically harmful ultraviolet radiation could reach heavily populated regions of the Northern Hemisphere.

The models further agreed that high atmospheric concentrations of chlorine could result in a potentially significant redistribution of ozone, with depletion in the upper stratosphere partially offset by increases in ozone at lower altitudes.[19] This development could have major, if not fully understood, implications for global short-term weather and long-term climate. In addition, the findings confirmed that CFCs themselves were thousands of times more powerful than carbon dioxide in their heat-trapping capability and therefore could significantly aggravate the greenhouse warming effect.[20]

The study indicated, moreover, that the ozone layer was threatened not only by CFCs 11 and 12, which had been the original focus of international scientific concern, but also by other fully halogenated alkanes, which included the related CFCs 113, 114, and 115 and two bromine compounds, halons 1211 and 1301.[21] All these chemicals shared the properties of long atmospheric lifetimes and high efficiency in triggering the catalytic reactions that destroy ozone, but they were not yet incorporated into the predictive models because of insufficient data. Demand for CFC 113 as a solvent in the electronics and other industries was growing rapidly. CFCs 114 and 115, though not at the time widely produced, could find use as refrigerants and plastic-foam-blowing agents in the event that only CFCs 11 and 12 were to be regulated. The halons, though manufactured in relatively small quantities, were increasingly important as fire extinguishants; moreover, molecule for molecule, halons had a substantially higher potency than CFCs for breaking down ozone. (See Table 2.1.)

Continuing Uncertainties

Although the theoretical understanding of ozone had progressed considerably since 1974, great uncertainties still remained as diplomats began in 1986 to debate the need for imposing international controls on CFCs. Thirty years of measurements had not demonstrated any statistically significant loss of total ozone. (The seasonal ozone loss over Antarctica, dis-

Table 2.1 Ozone-depleting and related substances: characteristics and uses. Ozone depletion potential (ODP), chlorine loading potential (CLP), and global warming potential (GWP) are index numbers developed by scientists, expressed in relation to CFC 11 (which is arbitrarily assigned a value of 1). The higher the value, the more environmentally detrimental the chemical. The number for ODP reflects the relative capacity of each chemical to reduce atmospheric ozone concentrations; for CLP, its capacity to increase atmospheric chlorine concentrations; and for GWP, its effectiveness as a heat-trapping gas.

Substances	Ozone depletion potential	Chlorine loading potential	Global warming potential	Atmospheric lifetime (years)	1986 world consumption (1000 metric tons)	Current and potential uses[a]
Substances controlled at Montreal						
CFC 11	1	1	1	60	411	A, PF, R, S
CFC 12	1	1.5	3	120	487	A, AC, PF, R
CFC 113	0.8	1.1	1.4	90	182	A, R, S
CFC 114	1	1.8	4	200	15	A, PF, R
CFC 115	0.6	2	7.5	400	15	A, R
Halon 1211	3	0	n.a.	25	18	FF
Halon 1301	10	0	n.a.	110	11	FF
Halon 2402	6	0	n.a.	n.a.	1	FF

Other substances						
CT	1.1	1	0.3	50	1,116	CF, P, S
MC	0.15	0.1	0.02	6	609	A, Ad, P, S
HCFC 22	0.05	0.14	0.4	15	140	A, AC, PF, R
HCFC 123	0.02	0.02	0.02	2	n.y.p.	A, PF, R, S
HCFC 124	0.02	0.04	0.1	7	n.y.p.	A, AC, PF, R
HCFC 141b	0.1	0.1	0.1	10	n.y.p.	A, PF, R, S
HCFC 142b	0.06	0.14	0.4	20	n.y.p.	A, AC, PF, R, S
HFC 125	0	0	0.6	28	n.y.p.	R
HFC 134a	0	0	0.3	16	n.y.p.	A, AC, PF, R
HFC 143a	0	0	0.7	41	n.y.p.	R
HFC 152a	0	0	0.03	2	n.y.p.	AC, R

Sources: Ozone depletion potentials, global warming potentials, and atmospheric lifetimes: WMO, *Scientific Assessment of Stratospheric Ozone: 1989*, Global Ozone Research and Monitoring Project, Report no. 20 (Geneva: 1989), vol. II, pp. 46, 300, 313, 395. World consumption: UNEP, *Economic Panel Report* (Nairobi, 1989), p. 14; Federal Republic of Germany, Federal Environmental Agency, *Responsibility Means Doing Without: How to Rescue the Ozone Layer* (Berlin, 1989), pp. 26–27.

Note: n.a. = not applicable; n.y.p. = not yet produced. Data are subject to continuing research and should therefore be considered approximate; figures are rounded when original sources vary or provide ranges.

a. Abbreviations for uses:
A = aerosols
AC = air conditioning
Ad = adhesives
CF = chemical feedstock
FF = fire fighting
P = pesticides
PF = plastic foams
R = refrigeration
S = solvents

cussed in the next section, was considered an anomaly at the time and could not be explained by the ozone-depletion theory.) Nor did the models predict that at existing rates of CFC emissions any global depletion of ozone would occur for at least the next two decades. In addition, not only was there no indication of increased levels of UV-B radiation reaching Earth's surface, but such measurements as existed appeared to show *reduced* radiation.[22]

The WMO/UNEP report noted that there were many gaps in understanding, inconsistencies among data sets, and "little overall support for the suggestion of a statistically significant trend" in global ozone levels.[23] A 1986 NASA study stated that "close examination . . . of models and measurements reveals several disturbing detailed disagreements . . . [that] limit our confidence in the predictive capability of these models."[24] Later in 1986 the U.S.-based World Resources Institute, though advocating CFC controls, had to admit that "definitive proof of changes in natural ozone levels and of the primary role of CFCs is still lacking."[25] In July 1987, practically on the eve of the final negotiating session in Montreal, NOAA concluded that the "scientific community currently is divided as to whether existing data on ozone trends provide sufficient evidence . . . that a chlorine-induced ozone destruction is occurring now."[26] Writing in 1989, Sherwood Rowland observed that "statistical evaluation through 1986 gave no indication of any trend in global ozone significantly different from no change at all."[27]

The Antarctic Ozone Hole

Too late for analysis under the WMO/UNEP assessment, British scientists in 1985 published astonishing findings based on a review of land-based measurements of stratospheric ozone made at their Halley Bay station in the Antarctic.[28] So unbelievable at first were these measurements that the scientists had delayed publication for nearly three years while they painstakingly rechecked and reviewed their data and the accuracy of their instruments. They finally concluded that ozone levels recorded during the Antarctic springtime (September–November) had fallen to about 50 percent lower than they had been in the 1960s. Although concentrations recovered by mid-November, the amount of the seasonal ozone loss had apparently accelerated sharply beginning in 1979. The "ozone hole" (that is, a portion of the stratosphere in which greatly diminished ozone levels were measured) had also expanded by 1985 to cover an area greater in size than the United States.

This unexpected revelation was quickly confirmed by Japanese and U.S. scientists rechecking their own data sets.[29] Interestingly, it was discovered that U.S. measuring satellites had not previously signaled the critical trend because their computers had been programmed automatically to reject ozone losses of this magnitude as anomalies far beyond the error range of existing predictive models.[30] After reanalyzing the satellite data, scientists also found that there had been no significant trend in ozone loss before the late 1970s. The precipitous decline had apparently been triggered by some atmospheric development at that time, which was not to be clearly understood until nearly three years later (see Chapter 9). The WMO/UNEP report guardedly commented that "these data indicate that some mechanism is at work in the cold southern polar night or polar twilight that is not generally included in models. This clearly warrants further investigation."[31]

The ozone hole did not, however, provide any clear signal for policymakers at that time. Scientists in 1986 and 1987 were far from certain that CFCs were involved in Antarctica. Nor could they confirm whether the hole was a localized phenomenon peculiar to unusual polar conditions or an ominous precursor of future ozone losses elsewhere over the planet.[32] Since ozone concentrations over Antarctica did recover after each springtime collapse, the phenomenon was contrary to known theory and did not conform to the global model predictions of gradual and pervasive long-term depletion. To add to the confusion, the downward trend was broken in late 1986—just before the start of the diplomatic negotiations—when Antarctic ozone concentrations actually improved over the previous year.

There were several credible theories, besides the role of CFCs, to account for the ozone hole, including polar winds, volcanic activity, and sunspots.[33] Even Rowland, writing in 1986 just before the treaty negotiations began, admitted that "the causes of the massive seasonal loss of ozone over Antarctica are not yet fully understood, and its implications for the ozone layer above the rest of the earth are also uncertain."[34]

A year later Rowland appeared to believe that the Antarctic revelation was the "driving force" behind the negotiations.[35] Other observers have also ascribed the success at Montreal to the "dread factor" of Antarctica.[36] But for those closest to the process, these judgments seem more a product of hindsight, overlooking how little was known about this phenomenon during the negotiations. Disturbing new evidence from a major Antarctic expedition of September–November 1987 was not, in actuality, completely analyzed and released until six months after the treaty was signed (see Chapter 9).

During July and August of 1987 some U.S. environmentalists debated whether to put pressure on UNEP to postpone the final protocol negotiating session until after the Antarctic expedition, in hopes that its results might influence governments to agree to more stringent CFC controls. However, the environmentalists worried that new data might prove inconclusive or again, as in 1986, show an ozone improvement. Such a result would undoubtedly have broken the momentum toward consensus, and the treaty negotiations would have become stalled in new bickering over the significance of Antarctica. The environmentalists finally decided that it was more prudent to take the "bird in hand" in September 1987 rather than risk further delay. Ironically, some of them later criticized the protocol for being too weak.

The hole over Antarctica did attract additional public attention to the ozone issue (though more in the United States than in Europe and Japan, where greater public pressure on governments was most needed). It may also have influenced some participants in the negotiations as evidence of the fragility of Earth's atmosphere. Significantly, however, Antarctica was never discussed at the negotiations, which were based solely on the global models. Even two months after Montreal, the U.S. Environmental Protection Agency had to conclude that "the Antarctic ozone hole cannot yet serve as a guide for policy decisions."[37]

Effects of Ozone Loss

Definitive evidence concerning harmful effects of ozone modification was even more sparse than proof of the atmospheric theories. Existing research, however, though tentative in many respects, did indicate a potential for extremely serious and wide-ranging damage to humans, animals, plants, and materials.

In June 1986, EPA sponsored with UNEP a weeklong international conference on risks to human health and the environment from ozone loss and climate change. This conference produced a large compendium of scientific papers, later supplemented by a multivolume EPA risk assessment, which together represented the most thorough study of the subject up to that time.[38] The scholarly work stimulated by EPA exerted a considerable influence on the subsequent diplomatic negotiations.

The link between ultraviolet radiation and skin cancer had been well established. However, contrary to scare stories in the U.S. media, there was "no evidence that a decrease in the ozone layer is responsible for the recent increase in the incidence of skin cancers."[39] Since such cancers take decades to develop, any decreases in ozone would have been too recent to account for the rising trend in skin cancer during the 1970s and 1980s. That trend was possibly attributable to other factors, such as changes in life-style resulting in more exposure to the sun—although this was also not scientifically proved.[40]

However, future ozone depletion would have serious consequences. EPA estimated that there could be over 150 million new cases of skin cancer in the United States alone among people currently alive or born by the year 2075, resulting in over 3 million deaths (with an uncertainty range of 1.5 to 4.5 million).[41] On the basis of the same parameters, EPA also projected 18 million additional eye cataract cases in the United States, many of which would result in blindness.

For other theoretical effects of excessive radiation, quantitative predictions proved difficult. For example, evidence from animal research indicated that UV-B could suppress the immune system. It was not possible, however, to determine the extent of increased human susceptibility to infectious diseases—even though this aspect was potentially very dangerous.[42]

Major damage to agriculture was also suspected. Laboratory tests indicated that some two-thirds of 200 plant species (including peas, cabbage, melons, and cotton) were sensitive to UV-B radiation, although this had not been confirmed under field conditions. The only existing long-term field studies were of soybeans, which did show substantial yield losses resulting from increased levels of UV-B.[43] Also extremely worrisome—but unquantified—was the potential impact on the productivity of fisheries, via possible disruption of the aquatic food chain caused by radiation damage to phytoplankton and other organisms living or reproducing near the ocean surface.[44]

Ultraviolet radiation was also implicated in costly accelerated weathering of polymers and in increased formation of low-level ozone (urban smog), injurious both to human health and to crops.

The potential effect of CFCs on global climate was related both to the redistribution of ozone at different altitudes and to the action of CFCs themselves as heat-trapping gases. Quantitative assessments were crude, but there was growing scientific consensus that greenhouse warming would have far-reaching implications for rainfall and agriculture, sea levels, and the survival of many animal and plant species whose habitat would be seriously modified.[45]

All of these possible effects were known to the negotiators of the Montreal Protocol, and they were never seriously contested. It was generally accepted that changes in the ozone layer would pose serious risks to human health and the environment. The point of contention among the participating governments was the extent of international action necessary to provide a reasonable degree of protection.

Spray Cans and Europolitics

3

Against a background of potentially great worldwide harm but considerable scientific uncertainty about whether that potential would be realized, the United States and the 12-nation European Community emerged as the principal protagonists in the diplomatic process that culminated in the Montreal Protocol.[1] Despite their shared political, economic, and environmental values, the United States and the EC disagreed over almost every issue at every step along the route to Montreal.

Different Approaches to Risk

U.S. and European official perspectives on ozone were conditioned by the relative influence and interest in the subject on the part of U.S. and Western European scientific communities, public opinion, and industrialists. A combination of factors disposed the United States toward enacting stronger measures to protect the ozone layer.

Symbolic of the difference between Europe and the United States in the importance accorded to this issue was the stratospheric ozone protection amendment to the U.S. Clean Air Act, passed by Congress in August 1977 in response to public reaction to the revelations of potential danger to the ozone layer. The amendment authorized the administrator of the U.S. Environmental Protection Agency to regulate "any substance . . . which in his judgment may reasonably be anticipated to affect the stratosphere, especially ozone in the stratosphere, if such effect may reasonably be anticipated to endanger public health or welfare."[2] This legal authority attempted to balance the scientific uncertainties with the risks of inaction. And it opted for a low threshold at which to justify governmental measures to confront dangers to the stratosphere.

Of critical significance in this legislation was the concept that was eventually to shape the U.S. position on international controls, the entire negotiating process, and the final treaty itself—namely, that a governing authority is not obligated, before regulating a particular substance, to prove conclusively either that it modifies the stratosphere or that the consequences are dangerous to health and the environment. All that is required is a standard of *reasonable expectation.*[3] This approach was unfamiliar at the time to most European officials, and it was destined to perplex EC negotiators throughout the subsequent international debate.

Even before this legislation was passed, however, EPA had proposed, under authority of the Toxic Substances Control Act, to prohibit by 1978 the use of CFCs as aerosol propellants in nonessential applications.[4] This ban affected nearly $3 billion worth of sales in a wide range of household and cosmetic products, from hair spray to furniture polish.[5] U.S. production of CFCs for aerosols fell by 95 percent, but substitutes and alternatives were soon on the market. The United States was followed in this action by Canada (a small producer) and by Norway and Sweden (nonproducing, importing countries). In effect, these governments determined that the benefits of CFC-propelled sprays were negligible when weighed against the potentially huge costs if the scientific theories proved correct.

Dominated by major CFC producers in France, Italy, and the United Kingdom, the European Community's approach to the hypothetical threat was quite different. Under heavy pressure from their domestic chemical and user industries and facing no counterpressure from public opinion or environmental groups, most EC member states, together with the EC Commission, decided on an essentially symbolic response.[6]

In 1977 the EC turned down a Dutch proposal to require labeling of spray cans containing CFCs.[7] A 1979 call by the West German Bundestag for a Community-wide CFC aerosol ban was rejected.[8] There was heated debate in the Community's European Parliament in 1979 after its Committee on the Environment, Public Health, and Consumer Protection proposed a 50 percent reduction in CFC aerosols by 1981 and a total ban, equivalent to the 1978 U.S. action, by 1983. The Parliament rejected its own committee's recommendation and instead endorsed a weaker EC Commission plan.[9]

The result was enactment in 1980 of a 30 percent cutback in CFC aerosol use from 1976 levels, not to take effect until the end of 1981. At the insistence of the U.K. government, a proposed requirement that governments provide the EC Commission with data on aerosols was dropped. In addition, the EC decreed that production capacity for CFCs 11 and 12 should not increase beyond its 1980 level.[10] These guidelines were supplemented in 1984 by voluntary engineering codes designed to conserve CFCs and to inhibit their emission into the atmosphere.[11]

These European measures seemed to most observers disingenuous. One European scholar, Markus Jachtenfuchs, characterized the 1980 EC decision as "reacting to American pressure" and observed that it "comprises only a minimum solution conceived out of the need to demonstrate to the United States that the EC was willing to act against depletion of the ozone layer."[12] The 30 percent reduction was a trivial target, since European sales of CFCs for aerosols had already declined by over 28 percent from their 1976 peak, mainly as a result of West German actions.[13] As a British environmentalist, Nigel Haigh, commented, "There is reason to believe that the figure of 30 percent was chosen because it was known that it could be achieved without creating too much difficulty for the industry."[14] Furthermore, the EC subsequently defined "production capacity" in a manner—24-hour continuous plant operation—that would actually allow output to increase by more than 60 percent from current levels. Another British analyst, David Pearce, noted that this capacity cap would not begin to affect EC production until the end of the century.[15]

It seems evident that the EC measures were dictated more by commercial than by environmental concerns. The EC Commission was sympathetic to industry arguments that strong controls on aerosols would impose hardships because of the substantial existing overcapacity and the allegedly large capital investment required to convert to the hydrocarbon propellants used by companies in the United States.[16] The producers also maintained that these substitutes were unacceptably inflammable and that tens of thousands of jobs would be jeopardized by a U.S.-style aerosol ban; it is worth noting that the West German government expressed skepticism about these claims.[17] In addition, all the EC "regulations" were actually implemented by voluntary agreements with the manufacturers. In sum, these were painless moves, fully supported by European industry, that gave an appearance of control while in reality permitting continued expansion.

The Economics of U.S.-EC Disagreement

The divergent U.S. and EC official actions were reflected in economic developments. The year of the CFC-ozone hypothesis, 1974, coincidentally represented a historic peak for CFC production and use, which had been growing at an extraordinary 13 percent average annual rate since 1960. The United States was by far the largest producer of CFCs 11 and 12 in 1974, with 46 percent of the reported world total of 813,000 metric tons.[18] All EC countries together accounted for 38 percent, with the Federal Republic of Germany the largest producer, followed by France and the United Kingdom (tied), then Italy, the Netherlands, and Spain.[19] Two years later, in 1976, the relative shares of the two areas had reversed themselves: EC production had continued to rise and accounted for 43 percent of the reported world total; U.S. output had dropped substantially and amounted to only 40 percent.

The United States never regained its market preeminence. In 1985, the year before protocol negotiations began, EC countries continued to dominate world output of CFCs 11 and 12, reaching over 45 percent. In contrast, total U.S. production (aerosols and nonaerosols) had fallen to nearly half its 1974 peak, and its share had dropped even further—to under 28 percent (see Figure 3.1). To be sure, the United States remained the largest CFC-producing and -consuming country and was also headquarters for the largest producing company: Du Pont accounted for about a quarter of world production, with factories in the United States, Canada, the Netherlands, Japan, and Latin America.

When the negotiations for the Montreal Protocol began in 1986, the United States accounted for an estimated 30 percent of world output of CFCs 11 and 12 and the other six major ozone-threatening CFCs and halons eventually covered, compared with the EC's estimated 43 to 45 percent. Other countries had increased their shares over the years, especially Japan (11 to 12 percent) and the Soviet Union (9 to 10 percent), with much smaller amounts produced by Canada, China, Australia, Brazil, Mexico, Argentina, Venezuela, and India.[20] (There were later reports of small existing or planned CFC facilities in Iraq, Israel, South Africa, South Korea, and one or more Eastern European countries.)

The salient economic feature of this period was clearly the new dominance of the EC. Relative to gross national product, the value of the EC's production of CFCs 11 and 12 was over 50 percent higher than that of the United States.[21] During the 1980s, CFC-propelled aerosols, which had virtually disappeared in the United States, still comprised over half of CFC 11 and 12 sales in the European Community. The EC was the almost unchallenged CFC supplier to the rest of the world, particularly in the growing markets of developing countries. EC exports had risen by 43 percent from 1976 to 1985, accounting for nearly one-third of its production;

France alone exported 40 percent of its output.[22] In contrast, the United States consumed almost all the CFCs that it produced.

Public Perceptions and Legislative Action

To a significant extent, official disagreements between the United States and the European Community on ozone policy reflected important disparities in public perceptions of the danger. These differences influenced both the politics and the economics of the issue.

The ozone-depletion theory had from the start captured the U.S. public's imagination. The subject was featured prominently in the media and in congressional debates and soon began to influence consumer behavior. Even before the 1978 aerosol ban, the U.S. market for spray cans had

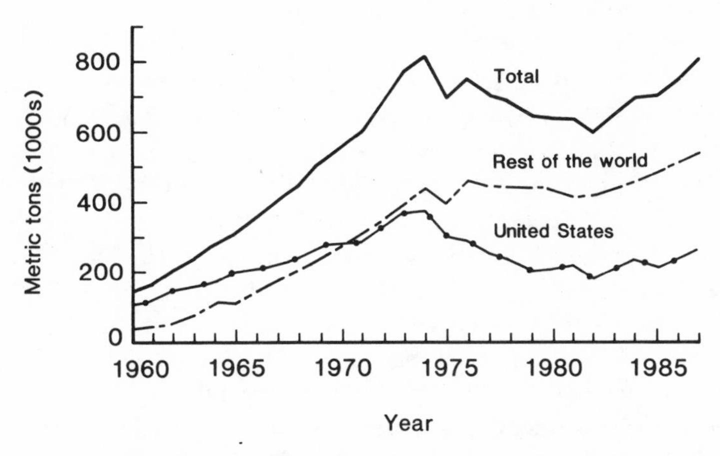

Figure 3.1 World production of CFCs 11 and 12, 1960–1987. (Rest-of-world data are primarily for the European Community and Japan, with small amounts for Canada, Australia, Brazil, Mexico, Argentina, and Venezuela. Data were unavailable for the Soviet Union, which in the 1970s accounted for 3–6 percent of the total and by the 1980s had risen to an estimated 9–10 percent. Also excluded are China and India, which by the mid-1980s together accounted for less than 2 percent.) *Sources:* Chemical Manufacturers Association, "Production, Sales, and Calculated Release of CFC-11 and CFC-12 through 1987," December 1988; U.S. International Trade Commission, *Synthetic Organic Chemicals,* annual series (Washington, D.C.).

fallen by nearly two-thirds because American consumers were acting on their environmental concerns.

In contrast, until mid-1987 there was no countervailing power in Europe to the chemical industry. It was not that the public in EC countries was less sensitive to environmental threats. Rather, Europeans long remained indifferent to the specific issue of ozone, which had a seemingly lower priority than such closer-to-home problems as acid rain, chemical spills, and the 1986 Chernobyl nuclear power plant accident.[23]

The U.S. media played an important role in keeping the issue before the American public through press and television coverage of the scientific theories and warnings over use of CFCs. After the diplomatic negotiations began in late 1986, media attention intensified; the ozone threat was featured in such widely circulated magazines as *Time* and *Sports Illustrated.*[24] Television news closely followed the course of the negotiations, and several special programs focused on the implications of ozone layer depletion.

The activities of American environmental organizations on the ozone issue also had no parallel in Europe.[25] These citizens' groups helped educate both the public and Congress by publishing studies, holding press conferences, and funding research.[26] One organization, the Natural Resources Defense Council (NRDC), filed suit against EPA in late 1984 to compel new U.S. controls. Under a court settlement in 1985, the EPA administrator agreed to publish in the *Federal Register* by May 1, 1987, a decision on whether or not additional domestic regulation was indicated according to the requirements of the Clean Air Act.[27] NRDC subsequently agreed to a postponement of this date pending completion of the international negotiations that had begun in the meantime.

The State Department encouraged U.S. environmental organizations to motivate their European counterparts to offset the influence of industry in the EC. In 1986 the World Resources Institute in Washington hosted a meeting on the ozone issue for European environmental groups (which State Department representatives scrupulously avoided attending). While diplomatic negotiations were under way in 1987, representatives of several U.S. nongovernmental organizations traveled to Europe and Japan to stimulate local environmental groups to take a stand on ozone layer protection.

A contrast in attitude also prevailed between the U.S. Congress and the parliaments of the major European producer countries, which, with the exception of the West German Bundestag, virtually ignored the subject of ozone. In the United States, formal hearings held soon after publication of the first ozone-depletion theory led to passage of ozone protection legislation in 1977.[28] Congress continued to follow the issue closely in the ensuing years. After the resumption of diplomatic negotiations in December 1986, both House and Senate endorsed the U.S. position on strong new international controls.[29] Legislation was introduced calling for unilateral U.S. measures in the event that the international negotiations failed. The bills included provisions for trade restrictions against nations not accepting their share of the common responsibility. (U.S. negotiators made certain that the implications of this threat were not lost on foreign governments, pointing out that there might be a price to pay for not joining in meaningful efforts to protect the ozone layer.)[30] More hearings were held by various House and Senate committees in the first half of 1987 to monitor the progress of the negotiations—and to sustain pressure on the negotiators.[31]

U.S. Scientific Leadership

Some observers believe that differences between Europe and the United States in public sensitivity to warnings concerning the ozone layer may have been related to the early space explorations and related American preeminence in stratospheric sciences. The sizable resources of NASA, NOAA, and similar institutions attracted more graduates, including many Europeans, into space science studies in the United States. When researchers at leading U.S. universities—California, Harvard, Michigan—voiced concern about events in the stratosphere, the media and the general public tended to pay attention.[32]

There was no equivalent in Europe to the NASA/NOAA research and satellite-monitoring initiatives on the ozone layer, nor to the series of National Academy of Sciences studies. As late as May 1988, 20 leading European scientists, meeting in Colorado, addressed a statement to the EC Commission noting that in stratospheric sciences "the gap between Europe and the United States is increasing rapidly." The scientists specifically called attention to inadequate funding and unavailability of satellites and state-of-the-art instruments. A few months later British scientists, in a public report, expressed disappointment that stratospheric ozone research had been excluded from the new environmental program of the EC Commission.[33]

Although there were a number of outstanding European stratospheric scientists (most of whom were invited by NASA to participate in the landmark WMO/UNEP international ozone project), they did not enjoy the same access to government policymakers as did their American colleagues. The U.S. government provided major funding for research on the ozone layer and then heeded the results. European scientists at the international meetings often found that they received a more attentive hearing from U.S. than from EC officials, and the American diplomats in turn profited from their scientific insights.

Indeed, during the period leading up to Montreal, it occasionally seemed as if some European governments either did not know where to get objective scientific advice or actually preferred to rely on industry lobbies for scientific information. After the 1985 discovery of the Antarctic ozone hole by British scientists, the U.K. government, citing budgetary constraints, refused to allot approximately $50,000 for additional instrumentation. It was a U.S. industry trade association—the Chemical Manufacturers Association—that ultimately provided the British Antarctic Survey with the needed funds, because U.S. industry wanted an early resolution of the scientific uncertainties. Two years later, shortly before the decisive 1987 negotiating session in Montreal, British scientists protested that a U.K. government press release had altered the tone of their scientific report to imply that CFCs were still thought to pose no great problem.[34]

Transatlantic Industrial Disputes

The chemical industries on the two sides of the Atlantic also developed different perspectives on the necessity for international regulations. U.S. companies seemed more willing than their European counterparts to adapt to changing scientific knowledge and public perceptions. This variance in outlook may have reflected long-standing cultural differences. Writing over 150 years ago, the French nobleman Alexis de Tocqueville observed: "I do not think, on the whole, that there is more selfishness among us [Europeans] than in America; the only difference is that there it is enlightened, here it is not. Each American knows when to sacrifice some of his private interests to save the rest; we want to save everything, and often we lose it all."[35]

Although Du Pont had declared in 1975 that restrictions on CFCs "would cause tremendous economic dislocation,"[36] producers in the United States, responding to consumer pressure, had moved to develop new propellants for spray cans even before the 1978 aerosol ban was introduced. These substitutes actually proved more economical than CFCs. Yet for several more years European chemical companies were able to persuade their governments that replacing CFCs in aerosols was unfeasible.

The strength of public concern confronted American industry with the threat of a patchwork of varying state regulations. Legislation against CFCs was introduced or passed in the mid-1970s in California, Michigan, Minnesota, New York, Oregon, and other states. As a result, U.S. industry not only became resigned to controls but even publicly favored federal regulations, which would at least be uniform and therefore less disruptive.[37]

U.S. producers—principally Du Pont, Allied, and Pennwalt—appeared increasingly more sensitive than the Europeans to the evolving science and the probable environmental risks from continued growth in CFC emissions. American firms seemed more worried about the impact of the ozone issue on their public image and long-term reputations. Their public pronouncements acknowledged that the ozone problem was at least potentially serious.

U.S. companies also resented the fact that their European rivals had achieved a competitive advantage in the late 1970s by blocking meaningful EC regulations. A 1980 EPA proposal to freeze nonaerosol uses met strong industry opposition on these grounds. U.S. industry repeatedly urged EPA to maintain a "level playing field" and to avoid recurrence of unilateral regulatory action by the United States that was not also binding on other major CFC-producing countries. Up until 1986, American producers—like their European counterparts—steadfastly opposed international regulation of CFCs.[38]

By 1986, however, many influential U.S. business leaders had begun to realize that the risk of ozone depletion would not go away, and moreover that production of CFCs had resumed an upward growth path. Pressures for new controls were once again mounting among environmental groups and in Congress. As the U.S. government intensified preparations for the resumption of international negotiations, it became apparent to American industry that the likelihood of an international control treaty could no longer be discounted.

In September 1986, less than three months before the scheduled start of diplomatic negotiations, the Alliance for Responsible CFC Policy, a coalition of about 500 U.S. producer and user companies, unexpectedly issued a policy statement supporting international regulation of CFCs. Alliance chairman Richard Barnett (who was also president of York, a major manufacturer of air conditioners) indicated in announcing the policy that the 1986 WMO/UNEP scientific assessment had influenced industry's evaluation of the situation. Although Barnett stressed that the CFC Alliance still did "not believe that the scientific information demonstrates any actual risk from current CFC use," he acknowledged that "large future growth in CFC emissions may contribute to significant ozone depletion in the latter half of the next century." In a conclusion that surprised many observers in the United States and abroad, Barnett declared that "large future increases in fully halogenated CFCs . . . would be unacceptable to future generations" and that "it would be inconsistent with the goals of this Alliance to ignore the potential for risk to those future generations."[39]

Reflecting differences within the business community, the statement did not endorse a specific set of controls. Rather, it recommended additional scientific research, conservation in CFC end-use, development of alternatives and substitutes for CFCs, and establishment of "a reasonable global limit on the future rate of growth of . . . CFC production capacity."[40] This position sounded modest enough, but it nevertheless represented an important breakthrough for a public stance by industry. American business representatives informally indicated that many companies would favor at least a freeze on CFC production and that others were prepared to accept substantial cuts—as long as adequate time was allowed for development of substitutes.

This policy change, which broke industry's transatlantic united front practically on the eve of international negotiations, was greeted with consternation in Europe. It contributed to obvious tensions between American and European corporate executives who attended the diplomatic negotiating sessions in the following months.[41] The British and French, in particular, were still irritated about a U.S.-European controversy in 1974 and 1975 concerning the Concorde supersonic jet. A major objection raised at that time to use of U.S. airports by the Anglo-French aircraft had been that direct infusions of nitrogen oxides into the stratosphere from the Concorde exhausts could damage the ozone layer.[42] Further research proved the U.S. argument defective, revealing, among other things, that the predicted chemical reactions would not occur at actual Concorde flying altitudes.

Some Europeans suspected that the United States was again using an ozone scare to cloak commercial motives.[43] They claimed to believe that U.S. companies had endorsed CFC controls in order to enter the profitable EC export markets with substitute products that they had secretly developed. This suspicion was unfounded. To the dismay of environmentalists, Du Pont admitted in 1986 that it had ceased research into alternatives for nonaerosol CFC uses five years earlier.[44] Alan Miller, a respected environmental scholar, concluded that before the protocol was negotiated, alternatives to CFCs, including recycling technologies, "had been explored and found wanting" by both producer and user companies.[45] Events after the signing of the Montreal Protocol conclusively demonstrated that there had been no secret substitutes on the shelf.[46]

For their part, the primary objectives of European companies, exemplified by France's Atochem, Britain's Imperial Chemical Industries, Italy's Montefluos, and West Germany's Hoechst, were to preserve market dominance and to avoid for as long as possible the costs of switching to alternative products.[47] Both industry and government officials felt that the Americans had been panicked into "over-hasty measures" in 1977 and therefore had only themselves to blame for any market losses.[48] Indeed, some Europeans hoped that delaying agreement on international controls might provoke impatient environmentalists in the United States to force a second round of U.S. unilateral regulations and thereby reinforce the EC's competitive position.

The EC Commission long based its position and tactics largely on the self-serving data and contentions of a few big companies. European industry views were echoed by official EC pronouncements. Both industry and EC Commission statements continually stressed the scientific uncertainties, the unfeasibility of substitutes, and the adverse effects that regulations would have on European living standards.[49] Epitomizing close industry-government linkages, company executives sometimes represented European governments on their official delegations to protocol negotiations—in contrast to their American counterparts, who were unofficial observers. In corridor conversations at the international meetings, some European industrialists and their government colleagues treated the ozone threat as trivial and were openly cynical about the objectives and prospects of the negotiations.

A former Dutch environment minister, Pieter Winsemius, and a member of the European Parliament, Beate Weber, were among those noting the unusually strong influence of EC industry on the Community's negotiating position in the case of ozone.[50] French observers maintained that "industry representatives substituted themselves [for] politicians in the decision making" on the ozone issue.[51] Another European analyst concluded that "in the last resort, it was the successful lobbying of two chemical companies [ICI and Atochem] at the national, and not at the European, level which determined the international position of the Community."[52]

The European Community in Transition

International discussions on possible harmonized measures to protect the ozone layer were strongly influenced by the political evolution of the European Community. The EC Commission was engaged at its Brussels headquarters in a historic effort to forge an economic and political union among the 12 member nations. Achieving such a union involved continuing pressure by the Commission to expand federal powers in areas that traditionally had been the prerogative of sovereign member states.

The Single European Act (article 130) in 1987 transferred formal responsibility for environmental matters from the member states to the EC Commission. In reality, however, in the words of one European scholar, the Commission "struggled with Member States to obtain an exclusive Community competence in external environment affairs," and there remained substantial ambiguities during the ozone negotiations concerning the actual extent of the EC Commission's authority.[53] U.K. officials testifying before the British House of Lords in 1988 insisted, for example, that "only limited decisions have been delegated to the Commission, and the Council [of Ministers] have the power to override [Commission] decisions."[54]

The Commission's centralized enforcement powers over member countries were incontestably weak. It relied on "dialogue" and on "encouraging" governments to enact its directives into national law.[55] As a 1989 European study noted, "Although broad in scope, Community environmental programs lack financial resources and manpower." The study also observed that "environmental activists—and EC officials as well—complain that environmental policies appear of marginal interest to most Member State governments."[56] The senior EC political-level official for the environment, Italy's Carlo Ripa di Meana, admitted in 1989 that levels of infringement by member states of some 150 existing articles of EC environmental legislation were "very high."[57]

In its approach to CFC regulation throughout the negotiations leading up to Montreal, the EC Commission seemed to many observers generally more concerned about European political and economic union than about the urgency of protecting the ozone layer. European environmentalists have "accused the Commission of endangering the adoption of the convention because of an issue which was more concerned with the Community's internal balance of power than [with] the protection of the ozone layer."[58] According to Ernst-Ulrich von Weizsäcker, a German political expert, "the motivation behind EC environmental policy . . . is not so much a common interest in environmental protection (even though this is, naturally, continually and solemnly asserted), but rather the harmonization of the conditions of production within the Common Market."[59]

Conflicts within the EC

Against this background, the EC was obliged to achieve internal consensus among 12 member governments before and during the international negotiations. Such consensus was hampered by the reality of deep divisions over the ozone issue within the Community. Belgium, Denmark, the Federal Republic of Germany, and the Netherlands were increasingly inclined toward strong CFC controls, but, of these, only Germany was a major producer.[60] The United Kingdom, France, and Italy, all large producers, resisted stringent measures. Greece, Ireland, Luxembourg, Portugal, and Spain did not even take part in most of the international negotiations; of these, Spain, which was home to CFC production facilities of French and West German affiliates, was known to strongly oppose new regulations.

In accord with its expanding federal role, the EC Commission insisted on being sole speaker for all 12 countries at the international negotiating table. It could not enforce this discipline, however, when the frustrations of individual delegations became too intense. French researchers characterized EC deliberations on ozone issues as marked by "harsh internal dissent."[61] Bitter behind-the-scenes struggles occurred at internal EC caucuses that preceded, and often accompanied, the negotiating sessions.

Resolution of these conflicts was complicated by the fact that the Council of Ministers (in this case environment ministers), which retained effective authority to determine the EC negotiating position, normally met only twice a year. Hence EC officials frequently declared that they were unable to respond to developments at the negotiations because they had no commonly agreed position. Although the Council may decide to act by "qualified majority," in practice unanimity is sought, a fact that can slow down the EC decision-making process and lead to convoluted compromises.[62] This situation facilitated efforts by commercial interests to exploit differences among member governments and thereby obstruct agreement. As Markus Jachtenfuchs noted, "The lack of a constructive mandate, or sometimes even of any position on issues to be negotiated, was a major difficulty for the Commission during the meetings."[63]

Another important factor was the EC presidency, which automatically rotates every six months among the member states. This office has substantial powers, including presiding over the Council and serving, alongside the EC Commission representative, as speaker for the Community at international meetings. European analysts have noted that "it was the reluctant attitude of the UK EC Presidency in the second half of 1986 that was largely responsible for the failure of attempts to seek compromises within the Council."[64] Progress in the ozone negotiations began only after Belgium replaced the United Kingdom in the EC presidency in January 1987. The United Kingdom remained in the EC "troika" (a three-country committee consisting of the past, present, and future presidents), which had an important influence in closed meetings of key delegation heads during the negotiations, but only until the presidency rotated again in July 1987. The U.K. delegate at Montreal, a Department of the Environment official who had played a prominent role throughout the negotiating process, was nonplussed when she showed up at the first of these crucial conclaves in September 1987 and was excluded by EC colleagues on these technical grounds.

The troika at Montreal in September 1987 comprised Belgium, Denmark, and the Federal Republic of Germany, all of which favored stringent controls. It is likely that this serendipitous constellation, in the right place at the right time, contributed to the EC's finally accepting considerably stronger measures than it had espoused throughout the negotiating process.

It is interesting to speculate on the extent to which the European Commission's attempt to pursue a common environmental policy helped or hindered international efforts to protect the ozone layer. Within the Brussels bureaucracy there was an understandable tendency to insist, in the words of EC chief negotiator Laurens Brinkhorst, that "the united EC approach [was] . . . environmentally beneficial."[65] On the other hand, Jachtenfuchs argued that "during all phases of the negotiations, the Commission played no leading role but had a more reactive and mediating part."[66] Even the top political-level EC official during the ozone negotiations, Commissioner of the Environment Stanley Clinton Davis, later conceded that "regrettably we were not able to take a firm lead as the United States did . . . [since] the United States has the advantages of a unified government approach."[67]

The United States indeed has powerful integrated resources and institutions at its disposal that were lacking in the European Community.[68] These include such strong federal agencies as EPA and the Food and Drug Administration, with far-reaching executive and enforcement mandates. The U.S. government also has centralized taxing and spending powers, industrial licensing authority, and a federal court system with precedence over state courts. All these factors are advantageous in implementing environmental protection, which involves reconciliation of complex economic, political, scientific, and public health considerations.

It is difficult to prove conclusively whether, during the ozone negotiations, the alliance of the United Kingdom and France in EC councils held back more progressively inclined member states—and thereby slowed movement on the wider international front—or whether it was the buildup of internal European pressure that finally forced the United Kingdom and other EC members to accept a treaty they might not have ratified at that time had they been totally free agents. Many observers inside and outside Europe might agree with Jachtenfuchs that "the restrictive position of the EC can be explained only by its specific internal decision-making procedures. Under the unanimity rule, two states strongly influenced by their respective chemical industries could force the others to join their position. This meant harmonization at the lowest common denominator."[69] EC policy certainly delayed ratification of the Montreal Protocol by several member states and nearly caused postponement of the protocol's entry into force (see Chapter 9).

However, once the EC accepted the treaty's obligations, it was able relatively swiftly to promulgate Community-wide regulations affecting all 12 member states. And as new scientific evidence in 1988 began to impose a greater uniformity of views among EC members on the threat to the ozone layer, the EC began to assume a leadership role. On balance, the struggles over ozone within the European Community were a microcosm of both the problems and the benefits of multilateral cooperation.

As the Community becomes a more fully integrated political and economic federation, it can be expected that the EC Commission's authority and powers will more closely correspond to those of the U.S. federal government and it will thus become, at least potentially, a more effective force for protection of the environment.

British Tenacity

Until mid-1987 the United Kingdom was the dominant voice on ozone in EC councils, where, as in many policymaking bodies, the party with the most strongly held convictions generally prevails. EC resolutions on CFCs often closely paralleled British policy documents.[70] Sources in the Community privately confirmed the statements of independent European observers that the United Kingdom consistently opposed strong international controls and attempted to delay an international accord.[71] Although France and Italy shared these sentiments, they were content for the British to be out in front.

Skepticism about the ozone theory and a minimizing of possible harmful effects marked the official U.K. position from 1975 through mid-1988.[72] At the time that the United States was extinguishing its CFC-aerosol industry, for example, the U.K. Department of the Environment concluded that "there appears to be no need for precipitate action on this issue." The U.K. government contended that "it is difficult to establish a direct quantitative relationship between decrease in the ozone column and the incidence of skin cancer."[73] An official U.K. submission to hearings called by the Council of Europe in Strasbourg observed that even "if the postulated ozone depletion did occur, it would result in increased exposure [to UV-B radiation] equivalent to a person moving from northern to southern England."[74] This argument missed the distinction between a voluntary move made with no knowledge of radiation consequences and the involuntary subjection of entire populations to known increases in UV-B. It also failed to consider that added radiation exposure would have a greater impact on people residing in less cloudy climes than England and would affect animal and plant species that could not relocate or otherwise compensate.

Because of the consistency of the British position, which certainly delayed and nearly prevented international agreement on meaningful CFC controls, it is worth examining factors that may have influenced the United Kingdom. The erroneous scientific hypotheses underlying earlier

U.S. objections to the Concorde may have influenced the British government to insist on greater scientific certainty before acting on this new version of an ozone alarm. Some observers have noted that Britons tend to perceive Americans as either overreacting to environmental dangers or using ecology to mask commercial motives. During the Concorde controversy, for example, the British journal *New Scientist* editorially expressed resentment against U.S. "environmental neocolonialism" and complained that U.S. economic power permitted it to "export its environmental conscience."[75]

Economics certainly played a role in the British position. CFCs were an important foreign exchange earner for the United Kingdom, which exported one-third of its CFC production. Aerosol products accounted for over 80 percent of the country's domestic use of CFCs in 1974, and even in 1986 they still amounted to over 60 percent.[76] One company with a major stake in the production of CFCs and halons—Imperial Chemical Industries (ICI)—influenced U.K. government policy throughout this period.[77] ICI was also the driving force in the European Council of Chemical Manufacturers' Federations, which was an active lobbyist in Brussels and a conspicuous presence during the negotiations.

In contrast to the vigorous activities of British industry, Parliament and environmental groups in the United Kingdom were essentially uninterested in the ozone issue during most of this time, and the media gave little attention to the subject. Not until early 1987 did the efforts of some U.S. environmentalists in the United Kingdom begin to pay off in the form of television interviews, press articles, and parliamentary questions about the government's negative policy. Indeed, these American private citizens were so successful that Her Majesty's Government in April 1987 asked the U.S. Department of State to restrain their activities.[78]

In any event, toward the end of the negotiations the Federal Republic of Germany—the largest CFC producer in Europe—began to assert its environmental judgments more forcefully within the European Community. Motivated by a combination of Green politics and growing concern over the scientific evidence, West Germany had by mid-1987 essentially replaced the United Kingdom as the primary influence on EC ozone policy.[79] Perhaps symbolically, at the final plenipotentiaries' session in Montreal, almost all the major EC actors—including Belgium, Denmark, France, the Federal Republic of Germany, Italy, and the Netherlands—dispatched ministerial- or ambassadorial-level representatives to sign the protocol. The one exception was the United Kingdom.

Prelude to Consensus

4

Any lasting solution to the ozone problem, which affects the entire world, must take place in a global context. Serving as a catalyst for such an outcome became the mission of the United Nations Environment Programme. Headquartered in Nairobi and operating on an annual budget of less than $40 million, UNEP proved indispensable to the process of arriving at an international consensus to protect the ozone layer.[1]

UNEP Starts the Process

Under the dynamic leadership of Mostafa Tolba, UNEP was active from the beginning in trying to sensitize governments and world public opinion about the danger to the ozone layer. UNEP made ozone protection a top priority in its program, and as early as September 1975 it funded a World Meteorological Organization technical conference on implications of the U.S. research. This meeting produced the first official statement of international scientific concern about CFCs.[2]

In March 1977, UNEP sponsored a policy meeting of governments and international agencies in Washington, D.C., which drafted a "World Plan of Action on the Ozone Layer." The plan of action recommended intensive international research and monitoring of the situation, and mandated to UNEP a central coordinating responsibility for promoting research and gathering relevant economic and scientific data.[3] UNEP established the Coordinating Committee on the Ozone Layer (CCOL), which in ensuing years undertook the important function of bringing together scientists from governments, industry, universities, and international agencies to assess the risks of ozone layer depletion. The CCOL produced periodic reports that served as valuable references for policymakers.

In April 1977 the United States hosted an intergovernmental meeting at which the question of international controls over CFCs was formally raised for the first time. Although a majority of participating governments favored at least a voluntary international agreement to eliminate nonessential aerosols, no consensus could be reached.[4]

The issue was considered again the following month at the annual meeting of the UNEP Governing Council in Nairobi, at which member governments consider and approve the organization's policies and programs. Canada, Finland, Norway, Sweden, and the United States sought at this point to expand UNEP's mandate beyond research to include consideration of international regulations, but this move was opposed by several countries as premature.[5]

In December 1978 the West German government hosted an informal meeting in Munich to reconsider this initiative. Participating governments were unable to reach agreement, even in principle, on quantitative goals for any CFC reduction. France and the United Kingdom blocked consensus in the EC, which voted formally as a single unit.[6]

The UNEP Governing Council approved a nonbinding resolution in April 1980 suggesting that CFC use be reduced but avoided setting any quantitative targets. The Governing Council further recommended that countries refrain from increasing production capacity, thus echoing the inconsequential EC policy.

In May 1981 the Governing Council finally responded to Tolba's urging and authorized UNEP to begin work toward an international agreement on protecting the ozone layer. However, most governments conceived of such a framework convention as covering only agreements on cooperative research and data collection, without actually imposing international controls. Later in 1981, UNEP convened a meeting of legal experts in Montevideo to consider aspects of such a global addition to the body of international law.[7]

The sense of urgency for new regulatory action had, however, diminished considerably. Changed understanding of the stratospheric chemical reactions, together with refinements in computer modeling, had led many observers to believe that the original ozone-depletion hypotheses might have been overstated (see Chapter 2). In addition, as a result of a worldwide economic slowdown and the drop in the U.S. aerosol spray market, total CFC production had declined. And in January 1981 a strongly antiregulatory administration had taken office in Washington; the new administrator of the U.S. Environmental Protection Agency, Anne Gorsuch, downplayed the likelihood of ozone depletion and cut back her agency's work in this area.

Notwithstanding these developments, Mostafa Tolba and UNEP persisted in their mission. It is no exaggeration to state that it was UNEP that kept the ozone issue alive at this stage.

Preliminary Jousting

In January 1982, UNEP convened representatives of 24 countries in Stockholm to launch the Ad Hoc Working Group of Legal and Technical Experts for the Preparation of a Global Framework Convention for the Protection of the Ozone Layer. The negotiations proved as cumbersome as the title. Because of the low priority now accorded to this issue by most governments, the deliberations stretched arduously through seven separate weeklong sessions over the next three years.

In 1983 Canada, Finland, Norway, Sweden, and Switzerland formed what became known as the Toronto Group, named after the city where these countries held their initial meeting, and introduced into the negotiations the idea of reducing CFC emissions. By this time Anne Gorsuch Burford had been replaced as EPA administrator by William Ruckelshaus, and midlevel EPA officials argued that it was inconsistent for the United States not to support internationally what it had already implemented domestically to protect the ozone layer. Thus, the U.S. government had another change of heart, and in late 1983 it joined the Toronto Group in proposing that the Ad Hoc Working Group develop a separate protocol containing international regulations, to be adopted simultaneously with the framework convention. Specifically, the Toronto Group advocated a worldwide ban on nonessential uses of CFCs in spray cans. It pointed out that American industry had already demonstrated that alternatives to CFC propellants were technically and economically feasible, although substitutes had not yet been found for other applications of CFCs.

A French study noted that at this point European industry "began to react violently."[8] The European Community made clear that it "was not even prepared to negotiate on any form of reduction of CFC production or use."[9] Followed by Japan and the Soviet Union, the EC firmly rejected the notion of an international regulatory regime.

By late 1984, however, the EC, perhaps apprehensive of adverse political repercussions from appearing so consistently intransigent on an environmental issue, adopted a new tactic. It proposed an alternative draft protocol text that would prohibit any additions to CFC production capacity. The EC argued, correctly, that without this prohibition, emissions reductions from an aerosol ban could eventually be nullified by uncontrolled future growth of CFCs for nonaerosol uses. The EC contended that its solution was environmentally sounder than the Toronto Group proposal, because it established an absolute upper limit on *all* worldwide CFC use. Moreover, the EC observed that what was a "nonessential" use for one country might not be for another. The French perfume industry, for example, seemed convinced that the alternatives to CFC propellants, though perhaps satisfactory for American insect spray, were unacceptable for their purposes.

The Toronto Group responded that the EC proposal was theoretically elegant but functionally ineffectual. The EC cap would permit its own industry, because of substantial excess capacity, to continue expanding at current rates for another 20 years (see Chapter 3). Millions of additional tons of CFCs would thus be released to do damage in the stratosphere. The cap would also lock in existing market shares and was therefore biased against countries with little or no surplus capacity; U.S. companies would be seriously disadvantaged under the EC formula.

In contrast, a worldwide ban on CFCs in aerosols would at one stroke reduce global CFC emissions by about one-third, yielding immediate benefit to the ozone layer. The U.S. chief delegate volunteered further that the Toronto Group proposal should not be considered the final word in international action; rather, it would buy time for science to provide clearer guidance to policymakers on the extent of additional controls that might prove necessary.

At the January 1985 negotiating session of the Ad Hoc Working Group, the United States called attention to a new theory by Harvard scientists that a sudden collapse of ozone concentrations might occur once the amount of chlorine in the stratosphere passed a certain threshold level. In his opening statement to the plenary, the U.S. representative declared that this new possibility heightened the urgency of effective short-term preventive action, and warned that "the margin of error between complacency and catastrophe is too small for comfort." [10] The EC, however, publicly dismissed these cautions as "scare-mongering," and the deadlock continued.[11]

Whatever the intrinsic logic of the respective proposals, it was evident that each of the two contending blocs was backing a protocol that would require no new controls for itself but considerable adjustment for the other. In an attempt to achieve at least some short-term emissions reductions, the Toronto Group then introduced a complicated "multi-options" approach, which combined aerosol reductions with a capacity cap.[12] This would have required at least some new controls by all parties. Unfortunately, analysis of the implications of the various options was impossible in the time remaining. Hence, this proposal foundered in the face of continued opposition from the EC, Japan, and the Soviet Union, reinforced by the indifference of other participating countries.

A British official, writing in 1985 on the negotiations for the Vienna Convention, noted that "not all countries . . . were prepared to be so sweeping or quick [as the Toronto Group] to condemn such useful chemical substances." He concluded that "their hesitance was vindicated" when newer model estimates showed reduced projections of future ozone depletion, and observed that, if the Toronto Group had insisted on adopting a control protocol in Vienna, "a sizeable proportion of the world's CFC-producing countries would simply have refused to have anything to do with it."[13]

The Vienna Convention

In March 1985 representatives of 43 nations, including 16 developing countries, convened in Vienna to complete work on the ozone convention. (See Appendix A for the text of the Vienna Convention.) Three industry organizations (the International Chamber of Commerce and two European federations) attended as observers; indicative of the environmental community's lack of interest in the ozone issue at this point was the nonparticipation by any environmental group.

By this time the Ad Hoc Working Group had not only achieved substantial agreement on a framework convention but also drafted all the elements of a protocol—with the crucial exception of the control provisions. The distinguished Austrian diplomat Winfried Lang, who presided over the plenipotentiary conference in Vienna and later chaired the negotiations leading to the Montreal Protocol, attributed the failure to agree on regulations to "industries in the EC who pretend that no alternatives are available to replace chlorofluorocarbons."[14]

With the stalemate over control strategies, the Vienna Convention for the Protection of the Ozone Layer was signed by 20 nations, plus the EC Commission. Signers included most of the major CFC-producing countries except Japan; the United Kingdom signed two months later. (For a list of countries signing and ratifying the Vienna Convention as of August 2, 1990, see Appendix D.)

The Vienna Convention was itself a considerable accomplishment. It represented the first effort of the international community formally to deal with an environmental danger before it erupted. The convention created a general obligation for nations to take "appropriate measures" to protect the ozone layer (although it made no effort to define such measures). It also established a mechanism for international cooperation in research, monitoring, and exchange of data on the state of the stratospheric ozone layer and on emissions and concentrations of CFCs and other relevant chemicals. These provisions were significant because, before Vienna, the Soviet Union and some other countries had declined to provide data on CFC production. Most important, the Vienna Convention established the framework for a future protocol to control ozone-modifying substances.

Symbolizing the reluctance of many participating governments at this stage, however, was the fact that nowhere did the Vienna Convention specifically identify any chemical as an ozone-depleting substance. The treatment of chemicals in the convention text was studiedly neutral as to why they were being recommended as subjects for observation and research. Chlorofluorocarbons were simply listed in an annex along with numerous other gases (some of which were known actually to promote *creation* of ozone in the stratosphere) as substances "thought to have the potential to modify the chemical and physical properties of the ozone layer."[15]

The United States and its allies introduced at the last moment in Vienna an unusual resolution, distinct from the convention itself. The resolution authorized UNEP to reopen diplomatic negotiations with a 1987 target for arriving at a legally binding control protocol. It further provided that, before the formal negotiations, UNEP would convene a workshop to develop a "more common understanding" of factors affecting the ozone layer, including costs and effects of possible control measures.[16] European industry opposed this initiative, and several governments were at best unenthusiastic. This resolution, which proved to be the springboard for the Montreal Protocol, was carried with the help of some countries in the

EC—Denmark, the Federal Republic of Germany, and the Netherlands—and with pressure from UNEP's Tolba, who was able to muster support from developing nations.

As incredible as it may seem in retrospect, it is worth noting that at the signing of the Vienna Convention in 1985, many governments were reluctant to designate UNEP as the official secretariat for the convention or any subsequent protocol. There was some feeling that the World Meteorological Organization, because of its more technical orientation, might be more appropriate. Despite (or perhaps because of) all the work that Tolba and his organization had done on the ozone issue, there was an unexpressed distrust in some quarters, and the secretariat question was by tacit consent left open at Vienna.[17]

A little-known episode at this time illustrated that the course of national policy seldom runs smoothly. Notwithstanding the departure of EPA Administrator Anne Gorsuch Burford, many officials in the Reagan administration, which was then beginning its second term, harbored philosophical objections to regulation and to United Nations multilateral powers. As the negotiators in Vienna were concluding their work, Under Secretary of State for Economic Affairs Allen Wallis recommended that Secretary of State George Shultz withhold authority for the U.S. delegation to sign the Vienna Convention. Wallis opposed the treaty precisely because he saw it as the forerunner to international regulation, and he warned that U.S. regulatory officials would view international agreements as a way to circumvent the administration's deregulation policies. Wallis also argued that coordination of research by international authorities might suppress independent scientific findings. Even though the ozone negotiations were far advanced, Wallis suggested that the Reagan administration's 1981 reversal of the U.S. position in the Law of the Sea negotiations provided a relevant precedent for a change of mind on ozone.

Frantic late-night transatlantic telephone calls by members of the U.S. delegation in Vienna, notably Assistant Secretary of State James Malone, alerted leading U.S. scientists and industrialists to the possibility that the United States might abandon the ozone convention at the last minute.[18] Even though industry representatives still strongly opposed controls, they had no ideological barriers against intensified international research: industry needed to know whether their activities could be dangerous, even if the eventual truth were to be unpleasant. Eleventh-hour private-sector interventions with the administration thus enabled the United States to join in signing the Vienna Convention. The incident curiously foreshadowed the more protracted attempt two years later by antiregulatory forces within the Reagan administration to overturn the U.S. position on the Montreal Protocol (see Chapter 5). In both cases, American industry ultimately backed the international agreement.

Watershed Workshops

The idea, launched in Vienna, of convening a special workshop significantly influenced the subsequent history of the ozone issue. In effect, having failed to agree on how to manage the risks, and being well aware of mutual suspicions about the commercial implications of any international controls, the contending parties agreed to step back and together reassess the whole situation.

A small international steering committee met twice in the ensuing months to plan this innovative process. The committee decided to divide the workshop into two weeklong sessions, and papers and studies were commissioned by various parties. The first workshop, sponsored by the EC in May 1986, was to examine CFC production and consumption trends, effects of existing regulations, and possible alternatives to CFCs. The second, sponsored by the United States in September 1986, would evaluate alternative regulatory strategies in terms of their implications for the environment, demand for CFCs, trade, equity, cost effectiveness, and ease of implementation.

UNEP invited all UN member governments to name both public- and private-sector experts to attend these workshops. The parties agreed to enter this exercise with open minds and no predetermined formal positions. There would be no official national delegations; participating government and UN officials, academics, industrialists, and environmentalists were all invited in their private capacities. Although financial assistance was offered to encourage participation by developing countries, there was a general lack of interest; China and India were among the few represented.

These informal workshops established a framework for the difficult formal negotiations ahead. The process was characterized by breaking down the problems into smaller components, developing consensus by incremental stages, and, as important as any other factor, establishing a degree of rapport and mutual confidence among future participants in the diplomatic negotiations.

The opening workshop in Rome was not propitious. Chaired by the

United Kingdom and dominated by European industry, the session nearly foundered in wrangling over CFC production and consumption trends. Although global output of CFCs 11 and 12 had fallen after 1974, a turnaround had occurred in 1983, and production appeared to be rising again at a 5 percent annual rate.[19] European industry representatives were reluctant, however, to admit the validity of any evidence that might be used to justify new controls. Their philosophy at the Rome workshop seemed to be that since projections necessarily entail uncertainty, nothing could usefully be said about the future. They maintained that the recent uptrend was a temporary phenomenon and argued vehemently against projecting any growth at all.

This contention seemed unrealistic to many participants. Worldwide demand for CFCs was increasing for many applications, including automobile air conditioning, rigid plastic-foam insulation, refrigeration, and industrial solvents. Indeed, the market was growing so rapidly that it was overcoming any inherent tendency for industry to delay investment in new plants because of uncertainty over possible future regulations. A Rand Corporation study estimated that worldwide CFC capacity had actually expanded by 17 percent from 1978 to 1985.[20]

Although data were incomplete, most analysts believed that, especially when CFC 113 was added in, the uptrend was incontestable. CFC 113, which had not even been seriously discussed during the negotiations leading up to the Vienna Convention, was finding new applications as a solvent in a growing range of industries. Du Pont estimated that worldwide production of CFC 113 had been increasing since the mid-1970s by more than 10 percent annually and that by 1985 it accounted for one-quarter of total CFC output. In 1984 the combined production of CFCs 11, 12, and 113 actually surpassed the peak year of 1974.[21]

The controversy in Rome over future CFC consumption and production trends was important. It was highly desirable that some agreement be achieved on this crucial input for the scientists' models of potential future ozone depletion. Most workshop participants considered it reasonable to project at least a 3 percent growth rate. However, consensus could not be reached, and the meeting ended with a compromise statement that future CFC production might grow anywhere from 0 to 5 percent annually.

Several weeks later Secretary of State Shultz designated the author to work with EPA in organizing the U.S. workshop and to serve as its chairman. The U.S. organizers decided to emphasize informality by establishing a retreat atmosphere at a conference center in rural Leesburg, Virginia, rather than holding the meeting in Washington. As a consequence, the participants were not subjected to big-city distractions; rather, under the inquisitive gaze of wild deer, they were together day in and day out. The shirt-sleeve working sessions were supplemented by evening barbecues, square dancing, a Southern-style plantation garden party, and bluegrass music—all of which helped to build personal relationships that were to carry over into the formal negotiations.

Whatever the auspicious influences of Indian summer in northern Virginia, the discussions proved to be refreshingly open and constructive. Substantively, much had been learned in the intervening months. High points had been release of the WMO/UNEP scientific assessment and the EPA/UNEP conference and publication on health and environmental effects, but there had also been several smaller events. EPA had issued a detailed "Stratospheric Ozone Protection Plan" in late 1985, and the agency was particularly active throughout 1986 in commissioning papers from academic experts and organizing workshops on various aspects of the ozone protection problem.[22]

Manifestations of growing scientific consensus—notwithstanding the considerable remaining uncertainties—undoubtedly contributed to greater open-mindedness among the participants at Leesburg. Many thoughtful papers were presented and discussed. Participants examined alternative control strategies in a relaxed and informal atmosphere, with the luxury of not being bound by official positions. Concerns over economic equity, problems of administration and monitoring, implications for developing nations, and similar issues were freely aired. It was possible to query, to interrupt, and to assume heretical positions just for the sake of argument, in a way that would have been impossible under the traditional etiquette of intergovernmental diplomacy. For the first time, for example, Soviet and Japanese representatives conceded that, at least personally, they could accept the need for international regulation. Government representatives from EC countries were generally less flexible in attitude; one U.K. official suggested that there was a long review period ahead before existing policies could be changed.[23] There was, however, a growing general belief that some kind of international regime was required, that past national positions would have to be modified, and that every country would have to make concessions.

Also for the first time, a new proposition was introduced into the discussions, termed at the time an "interim protocol." This concept implied that participating governments need not wait until they agreed on a definitive solution to the CFC problem. Rather, a treaty could be designed that would provide for periodic reassessments of the evolving science and that would contain built-in mechanisms for revising the controls if necessary. This idea facilitated future negotiations and proved to be a critical element of the eventual agreement.

In sum, those attending the Leesburg workshop shared a unique experience of collaborative problem solving. Most delegates seemed to agree when the chairman, in his concluding address, stated that the "spirit of Leesburg" offered a greater basis for optimism on the likelihood of an international accord than had previously been the case.[24]

Forging the U.S. Position

5

The weeks following the Leesburg workshop were marked by intensive activity in Washington. Although the administrator of the Environmental Protection Agency was responsible for promulgating and implementing U.S. regulations to protect the ozone layer, the Clean Air Act specifically required that "the President through the Secretary of State and the Assistant Secretary of State for Oceans and International Environmental and Scientific Affairs, shall negotiate multilateral treaties for these purposes."[1] In accordance with this mandate, Secretary of State George Shultz authorized the Bureau of Oceans and International Environmental and Scientific Affairs to coordinate an interagency process to develop a U.S. position and negotiating strategy. Under the leadership of the Department of State and EPA, and with significant scientific input from NASA and later from NOAA, approximately two dozen government agencies worked together in developing and promoting a U.S. position.

The Interagency Minuet

A series of interagency meetings, chaired by the State Department, began by examining the results of the various international workshops and studies. These meetings were later supplemented by smaller ad hoc working groups to deal with more specialized issues, such as trade provisions and legal questions. EPA staff prepared several comprehensive briefing books analyzing scientific, health, environmental, economic, and technological aspects of ozone protection. Scientists from NASA and NOAA interpreted the scientific findings and indicated their possible policy implications. International lawyers from the State Department contributed memoranda on legal precedents and issues. Experts from the Office of the U.S. Trade Representative provided documentation on foreign trade aspects of possible new regulation.

Throughout the deliberations, State Department and EPA representatives reported on their progress to a special senior-level working group of the White House Domestic Policy Council (DPC). The DPC was a cabinet-level body established by President Reagan to advise him on domestic matters and chaired by Attorney General Edwin Meese. Its formal membership consisted of the secretaries of Agriculture, Commerce, Defense, Education, Energy, Health and Human Services, Housing and Urban Development, the Interior, and the Treasury; the director of the Office of Management and Budget (OMB); the U.S. Trade Representative; and the Vice-President. For purposes of considering the ozone negotiations, the Secretary of State and the EPA administrator were added to the list. Several White House offices were also represented at the DPC working-group meetings, including the National Security Council, the Office of Science and Technology Policy, the Office of Policy Development, and the Council of Economic Advisers. The rationale for the major DPC involvement in the ozone issue, which began in the summer of 1986, was that the negotiation of an international treaty on chlorofluorocarbons and halons would require domestic implementing regulations.

As part of this interagency process, papers were circulated and discussed, special briefings were held for senior officials, and the outlines of a possible U.S. negotiating position began to emerge. Consistent with its Clean Air Act mandate, EPA had the primary responsibility for initiating possible regulatory options, which were commented on and modified by other participants in the course of the meetings.

From the discussions, it appeared obvious that reviving the 1985 U.S. position favoring a worldwide ban on aerosols would be a fruitless exercise, with no chance of acceptance by the EC and others. The agencies concluded that the United States should instead propose overall emissions limits, leaving the market to determine which particular end-uses were expendable or nonessential.

The interagency discussions also led to a consensus that in order to provide effective protection to stratospheric ozone, the protocol would have to go well beyond a freeze on emissions of CFCs 11 and 12 at current levels. There were several reasons for this conclusion. First, the science indicated that the other major ozone-depleting chemicals, in particular CFC 113 and the halons, would also need to be controlled. Second, significant reductions by parties to the protocol would be necessary to compensate for emissions from countries that might not join in the controls. Third, some increases for developing countries with low CFC-consumption levels would have to be permitted to encourage them to participate. Finally, unless industrialists were confronted with the prospect of a large drop in CFC production, it would not be profitable for them to renew research into alternatives; and the sooner substitutes were available, the smaller the likelihood that countries—especially the populous developing nations—would build new capacity in an obsolete CFC technology. Significantly, Du Pont had announced in 1986 that it believed it could develop substitutes in about five years but that "neither the marketplace nor regulatory policy . . . has provided the needed incentives" to justify the required investments.[2]

By November, agreement among the U.S. agencies on the principles of a U.S. negotiating position was incorporated in an administrative document known as a "circular 175," which provides the authority for the United States to undertake an international negotiation.[3] The circular 175 was drafted by the State Department and was formally cleared by the Departments of Commerce and Energy, the Council on Environmental Quality, EPA, NASA, NOAA, OMB, USTR, and the Domestic Policy Council (representing all other interested agencies). The document, which included a memorandum of law prepared by the State Department's legal adviser confirming the legal authority to negotiate, received final approval on November 28, 1986, from Under Secretary of State Allen Wallis—who less than two years earlier had tried to stop U.S. signing of the Vienna Convention.

The circular 175 established three principal elements for the U.S. negotiating position:

I. A near-term freeze on the combined emissions of the most ozone-depleting substances;
II. A long-term scheduled reduction of emissions of these chemicals down to the point of eliminating emissions from all but limited uses for which no substitutes are commercially available (such reduction could be as much as 95%), subject to III; and
III. Periodic review of the protocol provisions based upon regular assessment of the science. The review could remove or add chemicals, or change the schedule or the emission reduction target.[4]

The authorization to negotiate also included a draft protocol text, based on these principles, which the United States intended to submit to the other parties to the negotiations. The protocol text included a freeze at

1986 levels, followed by three "illustrative" reduction phases of 20 percent, 50 percent, and 95 percent, with the timing of these phases left open to negotiation. The concept of interim reduction stages was intended to promote technically feasible solutions by encouraging industry, knowing that more severe cutbacks were ahead, to tackle the easier problems first. Furthermore, the forces of competition, combined with the fixed schedule, should ensure that companies avoided deferring actions until the last minute. In addition, the timing of the reductions could be keyed to the results of the periodic technical reassessments. An important part of the U.S. text was a provision for trade-restricting measures to create incentives for nations to adhere to the protocol and simultaneously to protect U.S. industry against unfair competition from countries not submitting themselves to the treaty's requirements.

In discussions with foreign governments, environmental groups, and industry, the U.S. negotiators characterized this plan as a prudent and pragmatic insurance policy, providing a desirable safety margin against depletion of the ozone layer while scientific research continued. Its rationale was that acting now with reasonably strong controls would avoid much greater future costs both to industry and to society at large. U.S. officials maintained that the proposal would provide a degree of certainty for industrial planning. They also emphasized that the requirements should not be unrealistically stringent; adequate time should be allowed in the protocol for industry to develop alternatives.

The proposal was advanced as an illustrative approach rather than a firm position, and U.S. representatives were intentionally vague about details, particularly the crucial element of the timing of cutbacks. In reality, full agreement had not yet been reached within the U.S. government on the precise extent and timing of CFC reductions. Outside observers, including Congress and environmental groups, often chose to interpret "as much as 95 percent" as implying a firm commitment to virtual phaseout. However, U.S. officials insisted that although this was a long-term objective, the actual size and timing of reductions would have to be worked out during the international negotiations.[5]

As it happened, the delay in formulating a final position turned out to be a favorable factor in the diplomatic process. U.S. negotiators had to characterize their proposals as exploratory and to refer to principles and ranges rather than specifying precise details. This flexibility proved helpful in demonstrating to other countries the underlying reasonableness of the U.S. approach.

Predictably, this proposal completely satisfied neither environmental activists nor industry representatives: the environmentalists charged that it did not go far enough, while industrialists contended that it was excessively stringent. Nevertheless, the objections were not vehement. Both groups were in general sympathy with the government's basic goals, and they tended to cooperate with the U.S. official team throughout the negotiating process.

Diplomatic Strategy

The Department of State now designed and managed a multifaceted strategy to gain acceptance of the U.S. position by as many countries as possible. Over the next months about 60 U.S. embassies were regularly provided with talking points explaining the rationale behind the U.S. proposals, as well as with scientific and policy updates. Embassies were instructed to engage their host governments in a continuous dialogue to inform, influence, and demonstrate flexibility. A constant stream of cables between Washington and the embassies enabled the State Department to keep abreast of subtle changes in foreign attitudes and to provide new information responsive to other governments' concerns.

The U.S. negotiators coordinated these diplomatic initiatives closely with like-minded governments, particularly Canada, Finland, New Zealand, Norway, Sweden, and Switzerland. In addition, U.S. negotiators and embassies developed close relationships with countries in the EC that seemed sympathetic to the concept of stricter controls on CFCs, in particular the Federal Republic of Germany because of its volume of production. The United States also focused diplomatic attention on Japan, the Soviet Union, and key uncommitted developing-country governments.

Supplementing the embassy contacts, the chief U.S. negotiator led several missions to Western European capitals for consultations on both policy and science. When Belgium assumed the presidency of the European Community, the U.S. embassy in Brussels arranged for a meeting with the new Belgian environment minister, Miet Smet, who would be chairing the EC Council of Ministers during the first half of 1987. British officials had previously warned the United States that Belgium would support the Anglo-French antiregulatory position. However, after the Brussels meeting—which included parallel discussions between Belgian and U.S. scientists—Belgium became an active proponent of a strong treaty. It used its presidency during a crucial six-month period to help gain that end.

The U.S. diplomatic campaign placed great emphasis on the science.

American officials were careful to present a reasonable case and to distance themselves from exaggerated claims of imminent catastrophe. They contrasted the more responsible attitude of U.S. industry with the intransigence of the European chemical companies. American diplomats and scientists also deliberately did not stress the Antarctic situation, because of the major scientific uncertainties discussed in Chapter 2. They worried that linking the U.S. position with the ozone hole would risk its being undermined if that phenomenon turned out to be unrelated to chlorine. In the U.S. view, the case for stringent controls was strong enough without Antarctica.

The State Department's interim objective was to move the international political community gradually toward a consensus on the science and the risks. Enlisting support from NASA, EPA, and NOAA, the department organized a series of bilateral and multilateral scientific meetings to parallel the ongoing diplomatic negotiations. An offer to Soviet officials of bilateral cooperation in ozone and climate research was quickly accepted. The resultant pioneering U.S. scientific mission to Moscow in early 1987 helped to deepen Soviet understanding of the new findings and to change that country's previously adamant opposition to strong measures. A comparable scientific initiative and exchange with Japan in 1987 yielded similarly favorable results. U.S. officials in Washington and Tokyo sought appropriate opportunities to underscore in public and private the particular importance of mutual cooperation on this issue.

The media were an integral element of the diplomatic strategy. The U.S. government undertook major efforts to reach out to foreign public opinion, especially in Europe and Japan, to counteract the previously unopposed influence of commercial interests. Senior U.S. officials and scientists gave speeches, press conferences, and radio and television interviews in numerous foreign capitals. Using the advanced Worldnet telecommunications satellite technology of the U.S. Information Agency, Robert Watson of NASA, who had coordinated the WMO/UNEP science assessment, and the chief U.S. negotiator appeared together in a series of live televised question-and-answer sessions involving influential foreign participants in over 20 capitals in Europe, Latin America, and Japan. These programs, which went on for more than a year, attracted considerable foreign television, radio, and newspaper coverage.[6]

Highest-level personal diplomacy also played its role. Secretary of State Shultz and EPA Administrator Lee Thomas emphasized in personal contacts the seriousness of U.S. concern about the threat to the ozone layer.

At the June 1987 Venice economic summit of the seven major industrial democracies (Canada, France, the Federal Republic of Germany, Italy, Japan, the United Kingdom, and the United States, plus the EC Commission), an initiative by President Reagan succeeded in making protection of the ozone layer the first priority among environmental issues requiring common action.[7]

In addition to foreign diplomacy, the Department of State and EPA had to pay attention to important domestic constituencies. Senior officials testified before several different congressional committees on developments in the negotiations. Regular briefings were held for Congress, industry, and environmental groups. All these constituencies also sent their own observers to the negotiations, using this opportunity to press their often-opposing viewpoints on the U.S. negotiating team. With these observer representatives, the U.S. contingent at the international meetings was by far the largest, sometimes numbering over 30. Even the official U.S. delegation usually consisted of a dozen or more government agency representatives. As the next-largest national delegation had only four members, the Americans were a conspicuous presence. Added to the chief delegate's responsibility for conducting the actual negotiations were the tasks of maintaining a certain internal delegation discipline, at least among the U.S. agency representatives, and of keeping open a two-way flow of information with the observers from Congress, industry, and environmental groups.

It was essential to maintain good relations with the domestic groups and to reassure them that the U.S. negotiators would be fair and undogmatic. This was not always easy because of the differing points of view among the various parties. A few environmentalists demanded that the United States aim for a phaseout of CFCs within 5 to 10 years.[8] For their part, some industry representatives claimed they would be crippled by the apparent severity of the U.S. position on controls. Many in Congress expressed impatience with what they perceived as slow progress on the international front, and several hearings were held in the first half of 1987 to keep up pressure for a strong U.S. negotiating position.[9]

But at least one influential committee chairman, Representative John Dingell, Democrat of Michigan, felt things had been moving too rapidly. At a March 1987 hearing in the House of Representatives, he stated that he was "deeply concerned . . . [that] our chief negotiator, Ambassador Richard Benedick, and his EPA staff support . . . are negotiating on a seat-of-the-pants basis." Dingell represented a district heavily dependent on the fortunes of the U.S. automotive industry, which faced some discomfort because of reliance on CFCs in automobile and truck air conditioners. Dingell was unhappy that the chief U.S. negotiator, at the second international negotiating session in February 1987, "did not object to a draft bracketed protocol calling for a reduction in CFCs of a range of 10–50 percent in an unspecified number of years."[10] (In fact the United States, far from not objecting, had promoted that language; see Chapter 7.) Though generally endorsing an ozone treaty, Dingell underscored the scientific uncertainties and maintained that the U.S. position did not adequately reflect the adverse effect of the proposed controls on U.S. user industries. Less than two months later, however, Representative Dingell modified his stance: he accepted that there was "sufficient scientific consensus" for the protocol and stated that "we cannot afford to delay."[11]

All the U.S. domestic constituencies asked for assurances that the U.S. negotiators could not provide. Locked in a hard international negotiation, U.S. officials had to maintain an ambiguous public stance in order to make progress in the private bargaining. They could not indicate in advance to domestic observers what they might be prepared to compromise, lest they undermine their position vis-à-vis their foreign adversaries.

Antiregulatory Backlash: Hats and Sunglasses

While the U.S. diplomatic efforts and the international negotiations were proceeding, antiregulatory forces in the Reagan administration mounted a rearguard action in early 1987 to undermine the U.S. position on protecting the ozone layer. Although that policy had been approved by all the concerned U.S. agencies months earlier through the customary interagency process described above, some officials now raised complaints that it had not been endorsed at sufficiently high levels in certain agencies—specifically, that it had not been cleared with Secretary of the Interior Donald Hodel, Secretary of Commerce Malcolm Baldrige, and Presidential Science Adviser William Graham.

The reasons for this unexpected onslaught were uncertain. It was, of course, possible that the advisers to senior officials who would normally as a matter of principle have opposed environmental regulations simply had not been paying attention when the circular 175 had been developed and approved several months previously. They also may not have felt it necessary to expend great efforts at that time to contest the ambitious U.S.

position because the prospects for its success seemed so slight. It had been only one and a half years since the 1985 stalemate at Vienna over international controls on CFCs. Most observers, even those favoring strong controls, had believed that the possibility of a meaningful international agreement on ozone was doomed in advance, because of both the complexity of the subject and the rigid public opposition of the other major parties to the negotiation. In the fall of 1986—notwithstanding the "spirit of Leesburg"—not many wagers would have been placed on the likelihood that the powerful EC bloc, Japan, and the Soviet Union would ever agree to anything more stringent than a reduced rate of growth in CFCs, or at most a freeze in output.

However, as the U.S. diplomatic campaign gained both momentum overseas and support at home from influential voices in Congress, real reductions in production of CFCs and halons seemed increasingly less quixotic. Even as U.S. negotiators made progress in extracting concessions from the EC and winning over previously uncommitted countries, officials in the Departments of the Interior, Commerce, and Agriculture, together with the Office of Management and Budget, the Office of Science and Technology Policy, and parts of the White House staff, began in early 1987 to reopen basic questions about the scientific evidence and the possible damage to the U.S. economy from imposing additional CFC controls.

In March 1987 a senior OMB official, David Gibbons, established a special interagency working group outside the State Department process with the express purpose of reexamining the U.S. position. More volumes of data and analysis were generated for this review, primarily by EPA's seemingly inexhaustible staff under the direction of John Hoffman, an analyst who had persevered for several years in campaigning within EPA for stricter CFC regulations. Even these new studies, however, never seemed to provide the critics with sufficient detail or certainty. State Department and EPA officials, along with dozens of experts from NASA, NOAA, and the academic community, participated in a series of marathon, often late-night, briefings during early 1987 in an attempt to persuade the doubters of the reasonableness of the previously approved U.S. position.

For some, however, no persuasion seemed possible. Characteristic of the libertarian attitude on this issue was the argument that skin cancer was a "self-inflicted disease" attributable to personal life-style preferences, and therefore protection against excessive radiation was the responsibility of the individual, not the government. The flavor of the debate at this point was well captured in a letter written to Gibbons in April 1987 by one of the eminent U.S. experts on skin cancer, Margaret Kripke, chair of the Department of Immunology at the University of Texas System Cancer Center. After testifying at one of the OMB-organized interagency sessions, Kripke was so "troubled" by views expressed by Gibbons and others that she felt impelled to provide a written follow-up; she believed she had not responded adequately in person because she had been "caught unprepared for this line of reasoning." The letter found its way into the record of a congressional hearing. In it, Kripke expressed "discomfiture" at the characterization of skin cancer patients "as belonging to the leisure or sun-worshipping sets," noting that most of her clinic's patients were farmers, ranchers, and oil workers. She also rejected a suggestion that the upward trend in skin cancer was partly attributable to a "purely voluntary" movement of Americans southward to enjoy a more pleasant climate. Kripke observed that as a result of economic depression in the northern manufacturing states in the 1970s, augmented by high prices for heating fuel, many people "moved because of economic necessity, rather than because they want to play golf year round."[12]

In a similar vein, the following month brought an interview by Interior Secretary Hodel with the *Wall Street Journal*, in which he was quoted as saying: "People who don't stand out in the sun—it doesn't affect them."[13] On the same day the press reported that Hodel had, in a White House meeting, proposed a program of "personal protection" for individuals worried about ultraviolet radiation: reliance on broad-brimmed hats and sunglasses as a preferred alternative to more governmental regulation. The press accounts cited Hodel aides as the source of the leak; these aides apparently believed that publicizing this "common sense" alternative to costly controls would generate a backlash against overeager regulators at EPA and the State Department.[14]

Hodel's remarks, however, had the opposite effect: they were lampooned by editorials, by cartoons of fish and animals wearing sunglasses, and by environmentalists showing up at a Hodel press conference wearing cowboy hats and dark glasses, their faces smeared white with zinc oxide.[15] Hodel's proposals also aroused a storm in both houses of Congress, where senators and representatives rose to criticize them as "bizarre," "absurd," "startling," "laughable," and a "band-aid approach."[16] The "personal protection" plan was too much even for Representative Dingell, who earlier had shared some of Hodel's misgivings about the U.S.

position. In a June 1 letter to Hodel, made public in the *Congressional Record,* Dingell wrote: "I consider [the 'personal protection'] alternative to be detrimental to resolving a serious problem . . . I urge an international protocol in September, and urge you to abandon alternatives and to work toward that end."[17] In the face of such reactions, Hodel tried to backtrack and claimed that he had been misquoted, but the damage was done—opponents of the emerging treaty had been dealt a politically embarrassing blow from which they never recovered.[18]

A few days after the "hats and sunglasses" episode, the Senate debated a resolution on the U.S. negotiating position introduced by Senators Max Baucus, Democrat of Montana, and John Chafee, Republican of Rhode Island, with 33 cosponsors from both parties. The resolution urged President Reagan "to strongly endorse the United States' original [negotiating] position . . . [and] continue to seek aggressively . . . an immediate freeze . . . a prompt automatic reduction of not less than fifty percent . . . and the virtual elimination of [ozone-depleting] chemicals."[19] The views of the other side were summed up by Secretary Hodel and by Senator Steve Symms, Republican of Idaho.

Hodel maintained, in a letter reprinted in the *Congressional Record* during the debate, that his views had been "carelessly or deliberately" misrepresented by "those who, it appears, are determined to confine the President's options to those only of their crafting." He then raised a number of points suggesting that the U.S. negotiating position had been inadequately conceived. Noting that only a small proportion of United Nations member countries had been represented at the last negotiating session, Hodel implied that an international agreement would be "meaningless" because of nonparticipation by so many prospective producers or users of CFCs. Yet, even while insisting that near-universal participation was essential, he cast doubt on whether poorer nations should be "excused" from the same level of controls accepted by the United States. Hodel's letter also raised questions concerning the availability, cost, and safety of possible alternatives. He suggested that the treaty's proposed trade restrictions (designed to discourage nonparticipating countries from profiting) violated U.S. free trade principles. Finally, Hodel argued for delaying any controls beyond a freeze; he termed any future regulatory targets "highly questionable" in light of the uncertain scientific models.[20]

Senator Symms, in the June Senate debate, characterized the two opposing sides within the administration as represented by Secretary of State Shultz and EPA Administrator Thomas on the one hand and Interior

Secretary Hodel and the Office of Management and Budget on the other. For Symms, "the question is, is the sky really falling in on ozone? . . . One side is that the sky is falling in; the other side is that the ozone around the world really has not changed." He underscored the scientific uncertainty ("Are we . . . pushing forward based on some emotional viewpoint that is not completely verified yet?") and raised the specters of lost American jobs, lowered standards of living, and reduced international competitiveness if CFCs were to be regulated. Symms, like Hodel, also questioned the cost and availability of safe substitutes: "Will Americans have to make twice as many trips to the grocery store here because they will have less capable refrigeration units . . .?"[21]

The view of the ozone issue held by those opposing the U.S. negotiating position at this point was perhaps best summarized in Symms's statement to the Senate that "if people are going to insist on going out and exposing themselves to excessive sunshine, there is a limit to what can be done in terms of government . . . to try to protect people from . . . skin cancer."[22]

The Senate resolution passed by a vote of 80 to 2.

Cabinet Disarray

Notwithstanding the cartoons, an ideological dispute at high levels of government is no laughing matter. The OMB-led intensive reevaluation of the U.S. negotiating policy—which was widely reported in the press and which one congressional committee chairman characterized as "terrible" in its timing[23]—threatened the credibility of U.S. diplomacy at a critical point in the negotiations (see Chapter 6).

Critics also mounted efforts to change the makeup of the U.S. negotiating team, including ousting its leadership. The weekly journal *Human Events,* quoting Secretary Hodel's arguments against the protocol, charged that "officials at the State Department, led by chief negotiator Richard Benedick, and at the Environmental Protection Agency, have . . . push[ed] their own radical negotiating program for international controls on CFCs, and they have done so largely out of sight of the Administration."[24] And the *New York Times* quoted Hodel as accusing the chief negotiator of a conflict of interest based on his prospective assignment, following completion of the ozone negotiations, to The Conservation Foundation, an environmental think tank.[25]

On the scientific side, some critics had been uncomfortable with the presentations of NASA's Robert Watson. Therefore, senior Commerce Department aides enlisted an in-house expert, Daniel Albritton of NOAA's Aeronomy Laboratory in Boulder, Colorado, to present the science at subcabinet meetings. However, Albritton frustrated the antiregulatory faction when his meticulous reasoning and ingenious graphics corroborated the analyses of his NASA colleague.

The dispute over the U.S. position continued at several informal and formal meetings involving cabinet-level officials. During most of this period the State Department was represented by Deputy Secretary of State John Whitehead.[26] In meetings of the Domestic Policy Council, Whitehead and EPA Administrator Thomas argued the case for strong controls against a majority of skeptical department and agency heads.

A major break in the interagency debate came in the form of a cost-benefit study from the President's Council of Economic Advisers. The analysis concluded that, despite the scientific and economic uncertainties, the monetary benefits of preventing future deaths from skin cancer far outweighed costs of CFC controls as estimated either by industry or by EPA. This conclusion, which was based on the most conservative estimates and did not even attempt to quantify other potential benefits of preventing ozone layer depletion, dismayed the revisionists and helped sway some administration officials who had been watching the controversy from the sidelines.

Opponents of the emerging treaty responded with a new tactic: attempting to burden the U.S. negotiating position with unrealistic conditions. Such provisions, advanced ostensibly to make the protocol stronger and more effective, in actuality would have been impossible to negotiate. One example, particularly promoted by Science Adviser Graham and Interior Secretary Hodel, was that the United States should not ratify until it was assured that virtually all nations, including developing countries, would accept the treaty and adhere to stringent controls. Another was to demand unworkably strict verification and inspection systems. A third was to reintroduce into the negotiations the concept of nonessential uses: other major CFC producers should be required either to enact an aerosol ban or to grant the United States a credit for earlier emissions reductions resulting from its 1978 action; only after such "parity" was established should the United States agree to additional controls. Such new conditions, even though seemingly logical, would at this stage have damaged U.S. credibility as a negotiating partner and opened a Pandora's box of counterproposals by other countries; they would have had the effect of indefinitely stalling international agreement—which was the real objective of the antiregulatory faction. All these proposals had to be fought through the interagency process.

Interestingly, although some business circles had originally encouraged those U.S. officials attempting to modify the U.S. position and delay any treaty, at the critical moment major industry representatives declined to support them. Pragmatic industry executives traditionally paid more attention to the science and to public opinion than did those politicians and bureaucrats who opposed governmental regulation as a matter of principle. Besides, industry feared that a breakdown in the international negotiations might provoke a U.S. domestic political backlash that could lead to even more draconian controls. In the final analysis, industry preferred to face a stronger treaty, which would at least bind its foreign competitors, than unilateral U.S. controls with no treaty.

By this point the central element in the international negotiations, around which the controversy within the U.S. administration began to focus, was the timing and stringency of reductions in ozone-depleting chemicals (see Chapter 7). Most countries had by now agreed on an early freeze, to be followed within about four years by a 20 percent cutback. At the other end of the spectrum, the idea of a specific scheduled future reduction in the 95 percent range had received little support internationally, largely because of the remaining scientific uncertainties on the benefits of such a large cut; the United States, Canada, and others had fallen back to a position that the protocol should at least proclaim an "ultimate objective," based on the results of future scientific assessments, of eliminating all ozone-depleting substances.

But the critical issue was the proposal, which had been strongly backed by the United States, Canada, the Nordic nations, and others, that a second-stage reduction of about 30 percent (to make a total cutback of 50 percent) be firmly scheduled for approximately two to four years after the first reduction. This second phase was opposed by the EC, Japan, and the Soviet Union—and now also by several U.S. government agencies. A possible compromise—both internationally and in Washington—was to make the 30 percent second cut contingent on an affirmative decision by the parties to the protocol based on their future appraisal of the still-evolving science.

This idea seemed reasonable at first glance, and U.S. agencies that had gradually distanced themselves from the negative views of the treaty's critics found it appealing. As the final options paper on the U.S. negotiating position was being developed that June, only the State Department and EPA, among U.S. agencies, favored a "semiautomatic" 30 percent reduction—that is, a vote by the parties to the protocol would be required to overturn, rather than to enact, the second-stage cutback. This position, however, did have strong backing from Congress (as evidenced by the Baucus-Chafee Senate resolution) and from U.S. environmental organizations.

EPA and the State Department argued at the interagency meetings that a "20 percent protocol" would be a weak protocol, in some ways worse than no treaty at all because it would convey a false sense of having taken sufficient action. Since industry could easily achieve the 20 percent reduction with existing technologies for recycling and conservation (for example, eliminating use of plastic-foam egg and hamburger containers), only the additional 30 percent requirement would provide any real incentive for development of substitutes to CFCs. And this prospect had to be firm, rather than contingent, to remove planning uncertainty and to encourage industry to devote resources to constructive research rather than to fighting the possible reduction. A "20 percent protocol" would, in effect, invite a bitterly contested future renegotiation for the second-stage cut and thereby furnish an inadequate signal to the marketplace.

Presidential Approval

With the federal agencies still in disarray over this and other elements of the proposed treaty, it became clear that the issue would have to be decided at the top—by President Reagan. At this stage in the struggle, it may well have been the personal and behind-the-scenes interventions of Secretary of State Shultz that proved decisive in preventing a dramatic reversal of U.S. policy. Shultz made unmistakably clear to the president and other cabinet members his total support for the strong options.

The Department of State took the lead in formulating the arguments for the president in defense of a strong but effective U.S. negotiating position. However, preparation of the actual options paper itself was the responsibility not of the State Department but of the DPC staff, which was under heavy pressure from the Department of the Interior (Secretary Hodel was vice-chairman of the DPC), the Commerce Department, OMB, and White House Science Adviser William Graham. How the options were posed for the president became another point of contention when the State Department realized, on the eve of the crucial DPC meeting with President Reagan, that some of its carefully crafted language for its *own* proposed options had been altered at the last minute by White House drafters—distortions that slightly weakened the supporting argumentation and made the options seem less reasonable.

The meeting took place on June 18, with EPA Administrator Lee Thomas and Deputy Secretary of State John Whitehead defending a strong protocol. Acting for Shultz, Whitehead followed up with a letter that evening to White House Chief of Staff Howard Baker in order to counteract the subtle distortions in the DPC document. Whitehead's letter concisely summarized the State Department's position and annotated the rationale for each individual option now on the president's desk.

In its verbal and written communications with the White House, the State Department argued that, after months of promoting tough controls, including personal advocacy by the secretary of state and many U.S. ambassadors, a policy retreat at this stage would damage U.S. international credibility. Already several countries had expressed concern over press reports of possible backtracking by the United States, and the Federal Republic of Germany and Japan had pointedly noted, in private communiqués, that the evolution of their own positions had been strongly influenced by past U.S. representations.

The State Department also emphasized that the Clean Air Act provides a very low threshold for action to protect the ozone layer (see Chapter 3); thus, if the international agreement were too weak, the EPA might be obliged to regulate unilaterally—a less desirable outcome for both U.S. industry and consumers. Even if EPA proved reluctant to exercise authority on its own, pending congressional legislation and the NRDC court case could force strong regulatory action. The department called attention to the overwhelming bipartisan congressional support for a strong treaty, including "a prompt automatic reduction of not less than 50 percent."[27]

Finally, the State Department pointed out that the president was on the verge of scoring a major foreign policy and environmental success, which would be negated if administration proponents of a weak treaty had their way.

These arguments clearly had an impact on a White House that had become increasingly pragmatic, especially following Howard Baker's appointment as chief of staff. If the U.S. position were perceived to be weakened at this stage, a highly adverse political reaction could be anticipated both at home and abroad. Moreover, with a presidential election approaching in 1988, domestic opponents of the protocol had inadvertently made themselves into political lightning rods. As a result of Hodel's position on the ozone issue, for example, the *National Journal* opined in June that "the environmental movement . . . has found the Bad Guy it so sorely lacked in 1984—a Black Hat to personify resource exploitation, ecological despoliation, and environmental degradation." The article concluded that "the increasingly flamboyant Hodel . . . is their [environmentalists'] best election-year foil in a long time."[28]

With all these considerations, why should the White House go out of its way to reverse a position that had brought wide praise to the administration? Whatever the complex of reasons, President Reagan's decision on ozone, as on some other issues in the closing months of his incumbency (notably disarmament and relations with the Soviet Union), confounded his traditional ideological allies.

In a decision that was deliberately not publicized, the president overruled the antiregulatory group in his administration and accepted every one of the State Department options as interpreted in Whitehead's June 18 letter to Baker—including, to the surprise even of insiders, the critical provision for a virtually automatic 30 percent second-stage reduction. While unequivocally reaffirming the specific elements of U.S. policy on strong international controls, the presidential decision also allowed U.S. negotiators flexibility for the final phase of the negotiation. Some observers speculated that a primary reason for not revealing details of the decision was to forestall further public embarrassment to those in the Reagan administration who had opposed a strong protocol.

With this action President Reagan, who had, incidentally, undergone removal of two skin cancers in 1985 and one in 1987, probably became the world's first head of state to approve a national position for the ozone negotiations.

The Sequence of Negotiations

6

As the protocol negotiations began in December 1986, the only governments that had actually ratified the Vienna Convention were Toronto Group nations—Canada, Finland, Norway, Sweden, and the United States—plus the Soviet Union. Neither European Community countries nor Japan had yet ratified. This was not a propitious omen, since no regulatory protocol could become operational until the convention itself entered into force, and this required ratification by 20 governments. In a public statement in September 1986 the United Kingdom's Imperial Chemical Industries predicted that the ratification process for the Vienna Convention could take "several years."[1]

The Lineup of Countries

The negotiating parties appeared to be divided into three major camps, basically unchanged from the lineup some 20 months earlier at the 1985 Vienna conference. Following the 1986 Leesburg workshop, however, there were at least tentative hints of latent flexibility.

Officially, the European Community, negotiating as a bloc, followed the industry line and reflected the views of France, Italy, and the United Kingdom. The EC continued to advocate some form of production capacity cap. Because the scientific models indicated no significant ozone depletion for at least two decades, the EC Commission argued that there was time to delay actual production cuts and wait for more evidence. As late as November 1986, a U.K. environment official told the U.S. embassy in London that "most EC members do not attach much urgency to the ozone problem."[2] This perspective was shared by Japan and the Soviet Union during most of the negotiations.

Even at this late date the EC Commission still insisted that chlorofluorocarbons in aerosols could not readily be replaced "without serious loss of quality" and that "no fully satisfactory fluorocarbon alternatives would become available in the foreseeable future."[3] As in Vienna, the Commission still preferred that any international regulation be more symbolic than substantive, stating officially that the "status quo . . . should be maintained." An EC Council decision along these lines in late November meant, as a European observer noted, that "the EC was . . . not prepared to abandon its very restrictive 1980 position."[4] This decision elicited a formal protest from the West German government.[5]

In contrast, Canada, Finland, New Zealand, Norway, Sweden, Switzerland, and the United States publicly endorsed strong new controls. They argued that action had to be taken well before critical levels of chlorine accumulated. Because of the centurylong lifetimes of some CFCs, the atmosphere would unavoidably experience possibly dangerous future ozone depletion stemming from past and current production; the process could not be turned off suddenly. Despite the gaps in knowledge these governments were convinced that further delay would increase health and environmental risks to an unacceptable degree. They maintained that postponement of meaningful action could necessitate even more costly measures in the future.

A third group of active participants, including Australia, Austria, and a number of developing countries, were initially uncommitted, but as the negotiations progressed they moved toward favoring stringent regulations. During the negotiations Argentina, Brazil, Egypt, Kenya, and Venezuela played increasingly important roles representing the perspectives of developing nations.

Geneva: A Slow Start

Against this background the United Nations Environment Programme convened the opening weeklong negotiating round in Geneva in December 1986. As with all subsequent sessions, there were substantial international press and television coverage and many observers from industry, the U.S. Congress, and American environmental groups. Although UNEP had hoped for attendance by more than 50 governments, only 25 showed up. Nineteen were industrialized countries, and the other 6 were relatively advanced developing nations such as Mexico and Uruguay.

The negotiations began chaotically, with general and unfocused debate.

Canada, the Soviet Union, and the United States each proposed "illustrative" texts that were incompatible with one another. The U.S. text was the most comprehensive, covering not only control measures but also provisions for periodic assessments and adjustments, trade restrictions, and reporting.[6] Three Nordic nations—Finland, Norway, and Sweden—jointly offered an amendment to the U.S. text, calling for immediate cuts rather than an initial freeze. The Canadians proposed complex national emissions quotas based on a formula incorporating gross national product and population. The Soviets seemed unfamiliar with, and sharply critical of, the scientific rationale for new controls. They suggested national allocations based rather vaguely on population and CFC production capacity, with a complete exemption for developing countries.

For its part, consistent with its "status quo" decision, the EC Commission declared that it had no mandate to negotiate anything other than a cap on production capacity. EC and U.K. representatives then attempted to postpone the next round from the previously scheduled February 1987 date until April. They claimed that the EC could have no position until after its council of environment ministers convened in late March. When the plenary rejected this tactic, the heads of the U.K. and EC Commission delegations departed from Geneva prematurely.

Vienna: A Few Steps Forward

The following weeks were marked by intensive U.S. diplomatic activity to promote a serious long-term control strategy. The second session convened, as planned, in Vienna in late February 1987. In order to focus the debate, the U.S. delegation proposed that four separate working groups be established to deal individually with the issues of science, trade, developing countries, and control measures.

At Vienna there was growing evidence of evolution in the attitudes of many participating governments. Canada and the Nordic nations quietly abandoned their separate concepts and supported the proposed U.S. text outline. This format also gained backing from other countries, including Egypt, Mexico, New Zealand, and Switzerland. Japan and the Soviet Union remained enigmatic.

Important gaps separated the United States and the EC, however, on virtually every substantive issue. The EC stated that even a small reduction beyond a freeze would be very difficult to accept, although it could at least be considered. An informal EC proposal would have postponed even minimal 10 or 20 percent reductions in CFC emissions by nearly a decade. At a press conference in Vienna, and in separate interviews with Austrian and West German television and radio, the *New York Times,* and NBC, the chief U.S. negotiator characterized this position as "ridiculous" and "totally unacceptable" from the standpoint of protecting the ozone layer. He expressed disappointment at the slow pace of the talks and regret that "some participants at these negotiations seem to be concentrating more on short-term profits than on [a] common responsibility to conserve the environment for future generations."[7] Some observers commented three years later that this press statement, which received wide coverage, contributed importantly to building public pressure in Europe for a stronger treaty.

Divisions within the EC became increasingly evident at Vienna. The West German delegation, which included foreign ministry as well as environmental officials, was particularly unhappy with the EC Commission's delaying tactics. Two weeks after the Vienna meeting, Bonn termed the EC proposals "ineffective" and "insufficient" and threatened to initiate a unilateral ban on CFC aerosols regardless of what the rest of the Community might decide.[8]

An important step forward at Vienna, however, was the setting of a firm September date for the final plenipotentiaries' conference in Montreal. This both turned up the pressure and eradicated any lingering doubts or wishful thinking about the seriousness of the intent to push forward to a protocol.

Another significant result was the endorsement of a U.S. proposal to turn once more to the science. The scientists were asked to test on their models the future effects on ozone of alternative regulatory strategies that incorporated varying combinations of controlled substances and reduction schedules. A meeting of eight scientists from four countries, coordinated by Robert Watson of the U.S. National Aeronautics and Space Administration, was convened by UNEP in early April in Würzburg, West Germany. The results of this meeting had a decisive impact on the remainder of the negotiations.

Geneva Again: UNEP Takes a Stand

The number of participating governments rose to 33 at the third negotiating round, in Geneva in April 1987; of these, 11 were developing nations. UNEP Executive Director Mostafa Tolba, attending for the first time, set the tone with an impressive opening address. Referring to the work of the Würzburg meeting (described in Chapter 7), Tolba emphasized that "no longer can those who oppose action to regulate CFC releases hide behind scientific dissent." In the face of the potential of CFCs "to cause unmeasurable damage to our planet," he unequivocally placed UNEP behind tough international regulations.[9] From that point on, Tolba assumed a central role in the protocol negotiations, exerting his personal influence and his considerable authority as scientist and head of a UN organization.

Tolba organized in Geneva for the first time small closed meetings of key delegation heads, away from the formality of the large plenary sessions and the swarms of interested-party observers, and primarily focused on the crucial control measures. Working in secrecy on an unofficial text enabled representatives to negotiate with more flexibility, since they did not in this process commit their governments to any particular formulation. Membership in this informal but important body reflected a combination of a country's weight in the CFC market, interest in the ozone issue, and geographic balance. Tolba's group consisted of the heads of delegation of Canada, Japan, New Zealand, Norway, the Soviet Union, the United States, and the European Commission, plus Belgium, Denmark, and the United Kingdom (the EC presidential troika). The absence of any developing nation symbolized the South's lack of interest in details of the control measures; Tolba himself served, in effect, as representative of the developing world.

This group was able to produce an unofficial draft, labeled Tolba's personal text, at the end of the Geneva session. The draft represented considerable progress and increasingly resembled the original U.S. proposal. When Tolba's text was presented to the plenary, many governments expressed interest. Although the *New York Times* described the outcome as a "diplomatic triumph" for the U.S. State Department,[10] it was still too early for any party to give its official endorsement. Despite some movement by the EC, its differences with Canada, the Nordic states, and a growing number of other countries remained significant. On the basis of Tolba's text the press erroneously reported that an "agreement" had emerged at Geneva.[11]

Final Maneuvers

In late June 1987 Tolba reconvened his group of key delegation heads in Brussels to consider the controls and other major provisions. In July a small number of legal experts met in The Hague to analyze the entire protocol text as it had emerged from various working groups, in order to produce a relatively uncluttered and internally consistent draft for the final negotiating session in Montreal.

On the eve of the Brussels meeting, the chief U.S. negotiator received a secret cable containing President Reagan's instructions to maintain the strong U.S. negotiating position—but not to reveal details of the presidential decision. The injunction for secrecy inadvertently led to confusion among other negotiating partners, especially the European Community. As the U.S. domestic controversy was public knowledge, there was some uncertainty whether the strong positions being pressed by the U.S. delegate were still valid or merely the views of an individual who was about to be replaced in an administration shake-up.

As a consequence, the EC and U.K. showed unexpected intransigence at the meetings in both Brussels and The Hague. Indeed, during contentious legal discussions in The Hague, the EC and U.K. representatives tried to alter many clauses in Tolba's "personal text" from the Brussels negotiation. It was mainly Tolba's own forceful personality, together with the strong will of a young State Department lawyer, Deborah Kennedy, that prevented a watered-down draft protocol from being presented at the final negotiating session in September. Such an outcome would have effectively undone most of the convergence that had been achieved so laboriously in the previous weeks. Even so, the text contained numerous alternative formulations (in diplomatic parlance, "brackets") that required additional exhaustive negotiation in Montreal.[12]

During the summer, U.S. ozone diplomacy was again active throughout the world. The State Department invited EC Commission officials to Washington for bilateral consultations in an attempt to narrow the remaining large gaps. But the Commission politely declined, merely expressing optimism that compromises would be achieved at Montreal. According to informal sources in Brussels, the Commission, notwithstanding its assertion that it now possessed full authority to negotiate for all 12 member nations, was reluctant to be perceived by France and the United Kingdom as engaged bilaterally with the U.S. government at this critical juncture. In lieu of the thwarted U.S.-EC consultations, Deputy Secretary of State John Whitehead called in the EC ambassador to the United States from his vacation to make certain there was no residual misunderstanding of the U.S. position.

As part of the effort to reaffirm that the U.S. government remained steadfast, a series of cables went to U.S. embassies in many capitals. Seeking to muster support for the stronger alternatives in Tolba's text, embassy officials furnished other governments with detailed analyses of the various outstanding issues.

It is difficult to imagine the degree of tension and suspense among participants and observers as the Montreal conference approached. There was a sense that governments were entering uncharted territory. But the number and extent of issues still to be resolved in the putative closing round of a complicated international negotiation were staggering. Conflicts and uncertainties marked virtually every paragraph of the proposed protocol, from the central questions of which chemicals were to be controlled and the stringency of those controls to such crucial matters as trade restrictions and procedures for decision making and voting (see Chapter 7). The possible demands of developing countries, which would be attending the deliberations in significant numbers for the first time, were a totally unknown quantity. And claims raised by the EC Commission in July at the legal experts' meeting in The Hague concerning special treatment for members of "regional economic integration organizations" introduced a worrisome new element of contention. Characteristic of the midsummer mood was a message to the State Department from Tokyo, only a few weeks before the scheduled final negotiating round, that the Japanese government did not expect an agreement to emerge from Montreal.

Montreal: Agreement at Last

The parties reconvened in Montreal on September 8, 1987. The number of participating governments had now grown to over 60, of which more than half were developing countries. Scores of observers included, in sharp contrast to the 1985 conference on the Vienna Convention, representatives of many environmental organizations; industrial firms and associations were, as usual, also strongly in evidence. The international news media were well represented.

The first six days were devoted to attempts in various working groups to reach a greater convergence on the many bracketed portions of the protocol text. On September 14 the plenipotentiary conference was convened to complete the negotiations. Approximately 16 countries were represented at the ministerial or deputy ministerial level, including Administrator Lee Thomas of EPA, whose staff had labored so indomitably and effectively to reach this point.

With customary eloquence, Tolba in his opening address traced the long path of compromises that had led to this juncture. Recalling the negotiation of the U.S. Constitution, which had been signed 200 years earlier almost to the day, Tolba spoke of "horse trading," "frayed tempers," and "frustrations." In a pointed analogy to the current situation, he cited Benjamin Franklin's mixed feelings in signing: "I consent, Sir, to this Constitution because I expect no better." Tolba posed a question to the assembled ministers that clearly reflected his own frustrations: "Have we compromised so much that we have emasculated the agreement?" And he concluded, with mild optimism, that the controls provided in the accord would at least give the scientists, engineers, and planners "some time to think and act."[13]

But there was still to be considerable "horse trading" in Montreal. Although Tolba's text had gained widespread backing, major unresolved differences remained between the United States and the EC Commission. The summer of strong U.S. diplomatic efforts had paid off, however, and U.S. positions attracted support from a growing number of like-minded countries. Reflecting close consultations over the previous weeks in Tokyo and Washington, one of those governments was, to the surprise of many observers, Japan.

Tolba and the conference chairman, Austrian diplomat Winfried Lang, worked tirelessly throughout the eight days in closed meetings with key participants to hammer out the necessary compromises. There were the perhaps inevitable midnight theatrics. And a precedent-setting international accord was finally unveiled on September 16.

Vignettes from that day in Montreal reveal many departures from the normally staid diplomatic style and reflect the sense of history making that prevailed.[14] The atmosphere contrasted dramatically with the tentativeness and confusion that had marked the beginnings of the negotiations, only nine months earlier.

In the early morning of September 16, the head of the Japanese delegation exultantly waved a last-minute cable that contained his authorization to sign the protocol—a labor of dedication and burning the midnight oil in Tokyo. A young Chinese scientist, who had single-handedly and with distinction represented his country for the first time, promised before the plenary to do his "personal best" to persuade his government to join later. The chief Soviet delegate publicly endorsed the protocol, while apologetically explaining that his government could not sign in Montreal because a Russian-language text was not yet ready. The representative of Senegal, attending the ozone negotiations for the first time, confided that he had signed the protocol before actually receiving instructions from Dakar, because "it was the right thing to do." The Venezuelan delegation, which all morning had been suffering through the lengthy roll call of nations while anxiously awaiting word from Caracas, broke into cheers when a breathless messenger finally arrived with the authorizing cable—just in the nick of alphabetical time. And the expressions of weary relief on the faces of American industry observers contrasted with the stunned disbelief of many of their European counterparts.

It was an occasion that would long remain in the memories of those who were present. Mostafa Tolba as usual summed it up best in his closing address to the plenipotentiaries. Witnessing the fruition of 12 years of personal struggle, Tolba declared that the agreement had shown "that the environment can be a bridge between the worlds of East and West, and of North and South . . . As a scientist, I salute you: for with this agreement the worlds of science and public affairs have taken a step closer together . . . a union which must guide the affairs of the world into the next century." And, he concluded, prophetically, "This Protocol is a point of departure . . . the beginning of the real work to come."[15]

Points of Debate

7

The principal issues debated over the nine months from December 1986 to September 1987 can for analytical convenience be divided into eight categories:

1. what chemicals would be included
2. whether production or consumption of these substances would be controlled
3. the base year from which reductions would be calculated
4. the timing and size of cutbacks
5. how the treaty could enter into force and be revised, including the question of weighted voting
6. restrictions on trade with countries not participating in the protocol
7. treatment of developing countries with low levels of chlorofluorocarbon consumption
8. special provisions for the European Community

Against the general background history described in Chapter 6, each of these issues is examined here in more detail. (The text of the Montreal Protocol is reprinted in Appendix B. The protocol's key provisions are summarized in Table 13.1, pages 190–195.)

Chemical Coverage

Even though discussions before the Vienna Convention had focused only on CFCs 11 and 12, Canada, Norway, the United States, and others had come to insist, on the basis of evolving scientific understanding, that effective protection of the ozone layer would require *all* significant ozone-depleting substances to be controlled under the protocol. The original U.S. proposal in December 1986 had included CFCs 11, 12, and 113 and halons 1211 and 1301. U.S. negotiators had later added CFCs 114 and 115 to the list on the grounds that these ozone-destroying compounds would, if not restricted, simply be used in place of CFCs 11 and 12. Bolstered by new scientific findings, Norway proposed in Montreal to control a third halon, 2402.

Arguing more legalistically than scientifically, the EC long resisted going beyond CFCs 11 and 12.[1] The EC delegation head, Laurens Brinkhorst, charged in April 1987 that the Americans were complicating the negotiations by adding new chemicals. Moreover, the EC maintained that it lacked adequate data on the other compounds. EC views on this issue were exemplified in a paper—distributed to delegates by the French government under the letterhead of Atochem, the French CFC and halon producer—that stated: "He who grasps at too much loses everything, as they say in France."[2]

Japan was initially insistent that CFC 113 be excluded from control; it was an essential solvent in that country's expanding electronics industry. The EC and the Soviet Union were particularly reluctant to include the halons, which were important as fire extinguishants in sensitive defense- and space-related technologies and for which satisfactory substitutes were unavailable.

The turning point in this debate came from the conclusions of the scientific meeting held in Würzburg in April 1987.[3] The Würzburg analyses convincingly indicated that failure to regulate the rapidly growing CFC 113 and the extremely potent halons 1211 and 1301 would result in significant future ozone depletion—10 to 15 percent by the middle of the next century—even if the other CFCs were controlled.

In addition, the scientists had now agreed that each individual chemical could be assigned an index number representing its ozone-depleting potential (ODP). The ODP for a chemical was based on model calculations of its effect on the total column of atmospheric ozone. The value for a given chemical was calculated relative to CFC 11, which was arbitrarily assigned an ODP of 1.0. On the basis of this weighting system, the negotiators could craft a protocol provision that allowed substances to be treated for control purposes as a combined "basket" rather than individually. This formulation gave countries an incentive to impose greater reductions on substances that were relatively more harmful to the ozone layer, as well as those whose uses were less essential to them.

Because of the flexibility offered by this innovation, Japan dropped its opposition to including CFC 113 in the protocol. Japan could now take most of the required reductions from CFC 11 and 12 uses rather than having simultaneously to cut 113; moreover, Japan's technology for conserving and recycling 113, and therefore requiring less of the substance, was advancing rapidly.

Following the bilateral U.S.-Soviet scientific cooperation in the spring of 1987, the Soviet Union also came to accept the need for controlling a broader array of ozone-depleting chemicals, including halons. The European Community, however, steadfastly resisted imposing limits on halons. Even at Montreal, EC representatives held out for a nonbinding resolution that would have postponed any decision on halons until 1991.[4] They also claimed that they had no production data for halons and therefore proposed to move the base or reference year into the future. Both of these suggestions would have permitted unacceptable near-term increases in halon production. France and the United Kingdom were known to be major manufacturers of halons, but their companies had never publicly released output data.

After the Würzburg findings were thoroughly aired, there was no support at Montreal outside the European Community for narrow chemical coverage. The negotiators decided to include the five CFCs as one basket of controlled substances in the protocol and the three halons as another (protocol annex A). To be sure, the EC Commission, by using the leverage of withholding concessions on the more important CFCs, was able to exact a partial victory: it successfully delayed a freeze on halons until three years after the protocol entered into force, in contrast to the CFC freeze, which took effect in the first year.

Production versus Consumption

The issue of whether restrictions should be applied to the production or the consumption of controlled substances proved extremely difficult to resolve because of its commercial implications.[5]

The EC pushed hard for the production concept. European negotiators argued that it was administratively simpler to measure, and thereby to control, output, since there were only a small number of CFC- and halon-producing countries as opposed to thousands of consuming industries and countless points of consumption. The EC also feared that, if only consumption and not production were reduced, U.S. companies—which were currently operating at full capacity to meet domestic demand—would experience excess capacity as U.S. consumption was rolled back. With such excess capacity, American producers might be tempted to compete in the world's export markets, which were the nearly exclusive preserve of the EC.

Other governments, however, pointed out that focusing on production would convey inequitable power to existing producer countries, particularly the European Community, vis-à-vis other nations. For example, a production limit would essentially lock in the Europeans' foreign markets, which absorbed about one-third of total EC output. The only way that other producers could supply those markets would be to starve their own rising domestic consumption. Thus, EC exporters, with no viable competitors, would enjoy a monopoly reinforced by treaty obligations. Moreover, if continued growth in European demand for CFCs were at some point to tempt EC manufacturers to scale back on exports in order to satisfy their domestic consumers, CFC-importing countries, having neither recourse to other suppliers nor treaty authority to expand their own production, would have to bear a disproportionate share of reducing use of CFCs. Because of this vulnerability, CFC-importing countries might choose to remain outside the protocol and build their own CFC capacity. Controlling only production thus risked undermining the effectiveness of a protocol.

The reality was that the great majority of countries imported, rather than produced, CFCs. As early as the second negotiating round, Australia, Denmark (breaking ranks with the EC), Finland, New Zealand, Norway, Sweden, and the Soviet Union joined Canada and the United States in advocating a consumption-related formula.

The EC did have a valid argument regarding the impracticality of controlling multiple consumption points. To meet it, the United States and its allies came up with an alternative to either a pure production or pure consumption control: they proposed that a limit be placed on production *plus* imports *minus* exports. This surrogate for consumption, which was initially labeled "adjusted production," would satisfy the EC's concern about the difficulty of controlling consumption, since all three of its components were easy to measure. Relying on the market mechanism, it would eliminate any monopoly based on existing export positions.

If a CFC-importing country's traditional supplier raised prices excessively or cut back on exports, the importing nation could meet the shortfall either by substituting its own production or by turning to another CFC producer from among the protocol parties. In turn, a producing country could increase its own production (and exports) to meet such needs without having to reduce its domestic consumption.

Beginning in 1993, however, exports to nonparties could no longer be subtracted but would have to be counted against domestic consumption (article 3). This prospect would serve as an added incentive for importing countries to join the protocol, lest they lose access to suppliers. It would also stimulate current exporters (the EC) to encourage their customers to join.

For obvious reasons, European industry—and hence the EC Commission—did not welcome this proposal, and several times the negotiations threatened to break down. Sweden, the United States, and others continued to emphasize during the debates that, if only production were controlled, unfair benefits would be conferred on the EC while CFC-importing nations, especially the developing countries, would be at a disadvantage. Ultimately, the logic and equity of "adjusted production" proved to be compelling, and the opposition to a "production only" formula by a growing number of participants, including developing nations, became implacable. The EC Commission found itself isolated, and a solution was finally crafted at Montreal. Both consumption (defined as "adjusted production") and production would be frozen and then reduced according to an agreed-upon schedule (discussed below).

Two American officials, Robert Reinstein, of the Office of the U.S. Special Trade Representative, and James Losey, of EPA, devised a creative feature that provided additional needed flexibility: at each reduction stage, national production would be permitted to be slightly higher (10 to 15 percent) than national consumption. This concept, which was accepted at Montreal, would create excess supply in industrialized countries, allowing a limited expansion of exports to meet the legitimately expanding demand ("basic domestic needs") of developing countries that were parties to the protocol (article 2, paragraphs 1–4).

At Montreal, the Canadian delegation introduced an "industrial rationalization" clause that permitted this extra production allowance also to be used on behalf of smaller producer countries; if such a country could no longer produce efficiently because of small scale resulting from the required output reductions, it would be allowed to transfer its allowed production quota to another treaty party and to satisfy its needs by importing from that party (article 2, paragraph 5).

Some observers misinterpreted the differential between consumption and production and criticized it as a loophole in the treaty. It is therefore important to note that the differential applies only to individual countries; for all parties to the protocol considered together, total production cannot exceed total allowed consumption.

Base Year

The establishment of a suitable reference year for calculating the level of the freeze and the subsequent reductions was crucial. CFC production had started to rise rapidly again in 1983, and, if the protocol were to specify a reference year in the *future,* industry could be expected to expand output substantially to establish a higher basis from which subsequent cuts would be calculated.

Therefore, the United States, eventually joined by nearly every other actively participating country, would not compromise on its original proposal: the base year should be the one *preceding* expected completion of negotiations, or 1986. In an attempt to gain leverage for other issues, the EC Commission held out for 1990, claiming difficulties in obtaining export and import data for the earlier year.

But even after the Europeans finally conceded, the Soviet Union, which had not made an issue of the base year up to that point, dismayed the negotiators at Montreal by insisting that, unless 1990 were the base year, it could not join the protocol. Fortunately, it was discovered that the Soviet fixation on 1990 did not mask a scheme to maximize output over the next four years. Rather, the problem was the prior existence of a specific five-year plan moderately to expand CFC capacity. This plan, scheduled to end in 1990, was already under way and therefore apparently impossible to change under Soviet law. At the time, current per capita Soviet consumption was as low as that of most developing countries; even maximum fulfillment of the five-year plan would raise Soviet consumption to only about half the per capita level of the United States or the EC. Therefore, from the standpoint of equity, the Soviet stance was more reasonable than it had first appeared.

The breakthrough on this occurred in a way that typifies the element of chance in a negotiation. The base-year problem was being discussed at one of the small, informal meetings of delegation heads chaired by Tolba, which were the usual setting for difficult topics. Unlike the much larger formal plenary sessions, only English was spoken at these ad hoc negotiations, without benefit of simultaneous translation. For some time the chief U.S. negotiator had suspected that the Soviets were following the often rapid-fire discussion with difficulty. During a coffee break from the tense base-year debate, which at that point looked hopeless, a Soviet delegate happened to overhear the American speak to an Austrian colleague in the latter's native tongue. With evident relief, the Soviet exclaimed, "Ich habe's aber gar nicht gewusst, dass Sie Deutsch sprechen!" (I didn't know that you speak German). He was clearly more comfortable with German than with English, the mystery of 1990 unraveled, and thereafter informal communication on this and other issues improved dramatically.

Austrian chairman Winfried Lang and the U.S. representative were able, over lunch at a Montreal hotel, to design a "grandfather clause" to meet the specific situation of the Soviet Union. What later became article 2, paragraph 6, of the protocol permitted a country to add to its 1986 production base the output from any facilities that (a) were already specified in national legislation prior to January 1, 1987; (b) were under construction or contracted for prior to September 16, 1987 (the date of the protocol signing); (c) were completed by December 31, 1990; and (d) did not raise the country's annual CFC consumption above 0.5 kilogram per capita. This formulation was actually quite restrictive, and the total theoretically permissible expansion might never occur. Nevertheless, the Soviet negotiators accepted it with satisfaction, and the 1986 base year remained essentially inviolate.

Stringency and Timing of Reductions

For the general public, the most visible aspect of the protocol was the timing and extent of reductions. Not unexpectedly, this also turned out to be the single most contentious issue. Again, the European Community and the United States were the principal opponents.

The United States originally called for a freeze, to be followed by three phases of progressively more stringent reductions. The rationale for phased reductions was to provide milestones at which the parties could review the adequacy of the schedule on the basis of periodic scientific reassessments. In the U.S. draft text presented at the first session in December 1986, these cuts (which were bracketed to mark clearly their "illustrative" nature) were shown at 20 percent, 50 percent, and 95 percent of the base year.[6] During the early stages of negotiations the United States intentionally avoided specifying target dates for these reductions, stating only that they should occur soon enough to provide adequate protection for the ozone layer while also allowing time for industry to adapt.

EC representatives entered the negotiations hinting that they might consider lowering their existing capacity cap. This apparent concession, however, still implied some growth in production. At Vienna they progressed to a freeze, with a possible 10 to 20 percent reduction coming six to eight years after the protocol's entry into force. But the EC further weakened its proposal by insisting that any reduction step require an affirmative vote by at least two-thirds of the parties. The EC representatives at Vienna also refused to consider any additional specified reduction phases beyond the initial 20 percent.

Canada, Egypt, New Zealand, the Nordic states, Switzerland, the United States, and others rejected the EC proposal as completely inadequate. As a result of further negotiations in Vienna, an informal and nonbinding "chairman's text" was developed with the cooperation of Lang, specifying for the first time possible reductions of up to 50 percent. In the course of the maneuvering that led to Lang's text, the split in EC ranks became publicly evident. During heated debate, the EC Commission could not prevent Belgium, Denmark, the Federal Republic of Germany, and the Netherlands from openly joining the United States and others in defeating a U.K. attempt to weaken the terms and diminish the status of the chairman's document.

The Federal Republic of Germany, which had become increasingly concerned over the scientific evidence, was by now considering a 50 percent reduction regardless of what the rest of the EC might decide. Just before the Vienna meeting, Bonn had "urgently appealed" to the EC Commission and other EC members to support an "immediate freeze" on all CFCs, followed by an aerosol ban "within the shortest time interval."[7] (A European aerosol ban would in fact have resulted in a 50 percent reduction in EC members' total use of CFCs.)

The Würzburg scientific meeting in early April had major repercussions on this debate. The scientists demonstrated that serious damage to the ozone layer would occur under the weaker control options being considered by the negotiators.[8] The United States, through its embassies abroad, quickly disseminated the new scientific findings to a wide audience in foreign capitals. By the time of the Geneva meeting later in April, more governments had moved in the direction of favoring 50 percent cutbacks.

A momentum was building, and the structure of the crucial control provisions began increasingly to resemble the proposals presented by the United States at the first session in December 1986.

Mostafa Tolba's forceful intervention at the April 1987 session (described in Chapter 6) was another turning point in the negotiations. Tolba proposed a freeze by 1990, to be followed by successive 20 percent reductions every two years down to a complete phaseout in 10 years. New Zealand and Switzerland spoke out in favor of at least 50 percent reductions, with additional cuts to be dependent on future scientific assessments. The European Community, however, remained unwilling to move beyond 20 percent, and the Federal Republic of Germany again broke publicly with the EC Commission position in a carefully worded plenary statement that endorsed an eventual total emissions ban and expressed "great interest" in the New Zealand and Swiss proposals.[9]

Tolba then convened his informal consultative group of 10 delegation heads, meeting out of the limelight and away from the crowds of industry and environmentalist observers. (The group comprised Canada, Japan, New Zealand, Norway, the Soviet Union, the United States, and the European Commission, plus Belgium, Denmark, and the United Kingdom.) Though often heated, these discussions did produce a heavily bracketed text for article 2 (the control measures), for which Tolba assumed personal responsibility. This draft represented considerable progress, but it still enjoyed neither formal status nor approval by any government.

The Tolba text now specified a freeze within two years after the protocol's entry into force (EIF), followed by an automatic 20 percent reduction four years after EIF. Canada, New Zealand, Norway, and the United States had argued in the ad hoc meeting for a further automatic cut of at least 30 percent (for a total of 50 percent) within six years. But the EC, Japan, and the Soviet Union opposed this. The best that could be accomplished at this stage, even in an unofficial draft, was to present two options: the 30 percent second-phase reduction could occur either six years after EIF, but only if the parties took an affirmative decision; or eight years after EIF, unless the parties specifically voted to overturn it.[10] The EC, Japan, and the Soviet Union pointedly expressed reservations about the second-phase reduction in either format. They also allowed no mention of halons as controlled substances, and there was no resolution in this article 2 text of the base year or the production-consumption issues.

Tolba called a meeting of the same ad hoc group of key delegation heads on June 29–30 in Brussels, where he continued to urge further reconciliation of the differing views on article 2. This was the meeting for which the U.S. representative received the cabled presidential instructions referred to in Chapter 6. The EC Commission representative, Brinkhorst, not realizing that the dissension within the Reagan administration had been resolved, was unexpectedly inflexible. He branded the U.S. position as "extreme" and tried to enlist support from Japan and the Soviet Union for the "more moderate" European stance. The U.S. negotiators had hoped that some U.S.-EC convergence at the Brussels meeting might lead to substantial movement by the Japanese and the Soviets, particularly after months of U.S. diplomatic efforts in Tokyo and Moscow. But with the EC tactics ensuring that the two major blocs moved no closer to agreement, Japan and the Soviet Union remained discouragingly noncommittal.

Nevertheless, Tolba's draft continued to evolve, under pressure from Canada, New Zealand, Norway, and the United States, which reinforced his own evident preference for a stronger treaty. The CFC freeze was advanced from two years to one year after EIF. For the first time, halons appeared in the text, frozen at 1986 levels within three years after EIF; it was conceded that, at this stage, a freeze on halons would suffice because of their low volume. The automatic 20 percent reduction (of CFCs only) remained at four years after EIF. The controversial 30 percent second-stage CFC reduction was now scheduled for either 8 or 10 years after EIF, but the "prior approval" option disappeared and was replaced with a clause making this reduction also automatic unless it was reversed by a large majority. The text also included a provision that additional reductions could be imposed, following affirmative decisions by the parties, "with the objective of eventual elimination of these substances."[11] Canada, New Zealand, Norway, and the United States considered this clause important to provide a long-term planning signal to producers of ozone-depleting chemicals. Significantly, however, the EC representative continued to oppose both the 30 percent reduction and the concept of possible further cuts.

Notwithstanding the EC Commission's official intransigence, representatives of two of the three EC member countries present in Brussels (Belgium and Denmark) expressed personal optimism to the U.S. negotiator that by September the Community would support the Tolba text. This prediction turned out to be not quite on the mark.

Tolba's text for article 2 became the basis for the final negotiating session in Montreal and received support there from a wide range of countries. In addition to those that had previously publicly endorsed the 50 percent cut (Canada, Egypt, Finland, New Zealand, Norway, Sweden, Switzerland, and the United States), Tolba's proposal now received backing from Argentina, Australia, Japan, Mexico, Venezuela, and other countries that had been involved in consultations with the United States over the summer.

The EC Commission, after considerable haggling and several discordant caucuses with member-country delegations, finally accepted the semiautomatic 50 percent total cut. One price for this concession was the dilution of a clause prescribing an "ultimate objective of . . . elimination" of ozone-destroying chemicals and its displacement from article 2 in the protocol text to a less authoritative position in the treaty's preamble.

But there was still to be some unexpected drama over the timing of the reductions. At the eleventh hour, Brinkhorst, speaking for the EC, introduced an unusual interpretation of the phrase "within one year" as it appeared in the Tolba text.[12] He insisted that this meant that the freeze should not take place until the *second* year after entry into force, rather than during ("within") the first year. Similarly, the two reductions scheduled in the text for "within" the fourth and eighth years should, according to Brinkhorst's view, be recorded in the fifth and ninth years after EIF.

Although this interpretation seemed bizarre to most participants, the EC negotiator was adamant. After hours of stormy late-night argument, the only formula that proved acceptable was an awkward July–June compromise. To most observers, this meant simply that the EC had succeeded in delaying each round of control measures by six months.

The protocol text, as finally agreed upon, established a target date of January 1, 1989, for entry into force, with a freeze on CFCs at 1986 levels effective for the 12-month period beginning 7 months after EIF. The halons were frozen at 1986 levels for the 12-month period beginning three years after EIF. The automatic 20 percent CFC reduction would commence with the 12-month period beginning July 1, 1993, regardless of when the treaty entered into force. The additional 30 percent CFC reduction, unless reversed by a two-thirds majority of parties representing at least two-thirds of total consumption, would take effect with the 12-month period beginning July 1, 1998. Ironically, the chemical industry did not appreciate the EC representative's last-minute maneuver, because it presented companies with unwelcome problems of developing middle-of-year production data.

The *fixed* anchor dates for the two reductions were a significant feature of the treaty. They removed the temptation for any governments to stall the protocol's enactment in hopes of delaying the cutbacks (although the CFC and halon freezes could be affected by a delay in the treaty's entry into force). Simultaneously, they provided industry with firm dates upon which to base planning.

Crucial to the reduction timetable throughout the negotiations were the periodic scientific and economic assessments originally proposed by Canada and the United States, which would enable the parties to reexamine and, if necessary, revise any of the reduction steps according to the procedures described in the next section. There was never any controversy over this innovative mechanism (article 2, paragraph 9, and article 6).

Entry into Force, Revisions, and Voting

The interrelated issues of entry into force, revisions, and voting were not raised until near the end of the negotiating process. They proved, however, to be important elements of the final protocol.

The U.S. government had become increasingly disillusioned with one-nation/one-vote procedures in UN bodies, by which countries with substantial stakes in an issue could be overwhelmed on a majority vote. Precedents existed in some international treaties for weighted voting,[13] but the Vienna Convention for the Protection of the Ozone Layer had already established the one-vote principle for its protocols. Any attempt at this stage to overturn it would provoke objections from developing countries as well as others—for example, the Nordic states, which had otherwise supported U.S. positions on the ozone issue.

The U.S. solution was to introduce the idea of a two-step or qualified majority. Under this concept, certain actions could be undertaken only if they had the support of a certain number of countries, which together accounted for a certain proportion of total CFC consumption. It was necessary to use consumption rather than production for this purpose in order to allocate weight to all parties rather than only to producing countries. State Department lawyers advised that this two-step procedure was consistent with the Vienna Convention because each party retained one vote. At the same time, decisions could not be imposed unless the majority included parties most affected by the action.

This proposal, introduced by U.S. negotiators in June 1987 in Brussels, initially encountered resistance. However, explanations conveyed in the following weeks through U.S. embassies gradually persuaded other countries that the protocol would be ineffective unless major stakeholders had a corresponding influence in the collective decision-making process. The task at Montreal was to agree on the "certain proportion."

The domestic reexamination of the U.S. position discussed in Chapter 5 had revealed considerable distrust of the motivations of other large CFC-producing countries. There was concern that the United States could, in a situation analogous to its unilateral 1978 action, find itself bound to the obligations of an "international" protocol while its major competitors were not. As a legacy of the domestic debate, some U.S. agencies insisted on pushing for a proportion of consumption of 90 percent or higher as the trigger for entry into force and other actions. Such a provision could have proved a formula for delay, requiring unanimity among all four major blocs: the European Community, Japan, the Soviet Union, and the United States.

When the United States proposed this percentage at Montreal, the reaction was almost universally negative; only the Soviet Union supported 90 percent. Many observers feared that such a requirement could hold the treaty hostage to Japan or the Soviet Union, which might then weaken the protocol by extracting other concessions as the price for adherence. Smaller countries, including such U.S. allies as Finland and Norway, were irritated because they felt that such a high percentage denigrated their role. A few U.S. environmentalists charged in the press that this was a last-minute trick by the Reagan administration to prevent any international regulations from coming into effect.[14]

The 90 percent opening position did represent a face-saving gesture to the hardliners in the U.S. administration who had lost their battle to block regulations. It also, however, highlighted to all governments the importance of broad participation and of consensus among major stakeholders. The State Department negotiators had to maintain this position for a decent interval, absorbing criticism from both allied governments and environmentalists, in order to remove any lingering doubts in Washington that the position was fundamentally untenable.

An inevitable, and reasonable, compromise was struck in Montreal, providing that entry into force would require ratification by at least 11 parties, together constituting at least two-thirds of estimated global consumption of controlled substances as of 1986 (article 16). In effect, to become binding the protocol would have to be ratified by the United States and at least four of the six other large consumer countries (France, the Federal Republic of Germany, Italy, Japan, the Soviet Union, and the United Kingdom), or by the United States and the EC as a unit. Most observers believed that this would provide a sufficient critical mass to increase the pressure on any potential large holdouts to join the treaty.

The qualified-majority procedure was also applied to other decisions under the protocol. Cancellation of the 30 percent second-step reduction was intentionally made difficult, in part to discourage any die-hard industrial groups from expending effort on overturning the provision rather than on planning for a virtually certain cutback. To void the reduction, the treaty required agreement by two-thirds of the parties that together accounted for two-thirds of total consumption of all parties (article 2, paragraph 4). At a minimum, both the EC and the United States would have to concur that the science had become unexpectedly optimistic and that the second reduction was consequently no longer necessary.

On the other hand, the threshold for a positive decision to undertake additional reductions beyond 50 percent was made less demanding, and thus represented a victory for those governments favoring a relatively stronger protocol. Such action would require a two-thirds majority of countries representing at least 50 percent of total consumption of all parties (article 2, paragraph 9). Thus, if a sufficiently large consensus developed for reductions beyond 50 percent, neither the United States nor the EC alone could block it, although both together could. The protocol was so designed that such future changes in the stringency and timing of reductions of already controlled substances would be considered "adjustments" to the provisions and therefore binding on *all* parties, even those that had not voted with the majority. An international committee of legal experts later hailed this concept as a "great novelty in international environmental law."[15] The rationale for the binding feature was that these adjustments applied to known chemicals.

In contrast, however, adding future controls on new chemicals was terra incognita. Whether such actions should be fully binding for all parties was a problematic issue for some governments, which were ultimately not prepared to subordinate their decision to some future international vote. Accordingly, the negotiators at Montreal determined that the addition of new chemicals and controls would constitute an "amendment" to the protocol and therefore would be governed by article 9 of the Vienna Convention. Thus, although such a decision would require only a simple two-thirds majority vote, it would enter into force only after at least two-thirds of the parties formally ratified it, and would be binding only for those parties that ratified (article 2, paragraph 10).

Trade Restrictions

At the first session in Geneva in December 1986, the United States offered specific proposals to restrict trade in controlled substances with nonparties. The objective of such restrictions was to stimulate as many nations as possible to participate in the protocol, by preventing nonparticipating countries from enjoying competitive advantages and by discouraging the movement of CFC production facilities to such countries. These provisions were critical, since they constituted in effect the only enforcement mechanism in the protocol. Yet the trade issue also proved to be a complex and contentious subject, and trade working groups debated exhaustively at each negotiating session.

Initially, the EC proposed only that parties "jointly study the feasibility of restricting imports."[16] At the April 1987 Geneva session, the EC Commission and U.K. representatives blocked further discussion of trade until a legal expert from the General Agreement on Tariffs and Trade (GATT) secretariat could advise on the permissibility of such measures under existing international law. This tactic backfired when the expert affirmed that trade restrictions were allowable, under article XX of GATT, as they could be considered "necessary to protect human, animal, or plant life or health" and "relating to conservation or exhaustion of exhaustible natural resources."[17] From this point on, a consensus inexorably grew that such measures were indispensable to the protocol's effectiveness.

As a strong incentive for countries to ratify the protocol, the United States initially proposed an outright ban on CFC exports to nonparties. EC countries responded that this measure would unfairly affect them, as they had no assurance that their current customers would join. The eventual compromise provided that, after January 1, 1993, exports to nonparties could not be subtracted from a party's production in calculating its consumption level (article 3). Thus, if EC nations wanted to continue exporting to any customers that had not joined the protocol within four years, they would have to reduce their own domestic consumption to do so. Since such a move was unlikely, there would be pressure on importing countries to join the protocol in order to maintain their supply.

With respect to imports from nonparties, the final text banned the import of *bulk* substances within one year of entry into force (article 4, paragraph 1). In addition, U.S. negotiators had campaigned hard for restrictions on imports from nonparties of products *containing* or *produced with* any controlled substances. One rationale for such limits was to provide another incentive for potential holdouts to join the protocol, lest they lose their markets (for example, Asian electronic products using CFC 113 as a circuit cleaner). Another was that unless such products were controlled, producers could be tempted to shift CFC-manufacturing facilities to pollution havens offshore; the original producing country could then import the products without being itself accountable for the related CFC emissions. Such actions would nullify benefits to the ozone layer and also impede development of non-ozone-depleting chemicals.

These U.S. proposals encountered strong resistance from the United Kingdom and the EC Commission throughout the negotiations. Compromises were finally reached only in Montreal. A ban on products *containing* the controlled substances would come into effect four years after EIF, based on a product list to be developed by the parties (article 4, paragraph 3); such a list would probably include such items as refrigerators and air conditioners. Within five years the parties would "determine the feasibility of banning or restricting" imports from nonparties of products *produced with* (but not containing) controlled substances and would elaborate a list of such products—for example, many electronics items employing CFC 113 to clean microchips (article 4, paragraph 4). Both of these provisions, however, would be binding only on parties that concurred in the product lists. Finally, parties to the treaty agreed to "discourage" export of technology to nonparties for producing or utilizing the controlled chemicals, and to "refrain from providing" new financial aid (including guarantees or insurance programs) for such purposes; an exception was made for technologies that could contribute to emissions reductions, including containment, recycling, or destruction of controlled substances (article 4, paragraphs 5,6,7).

Although these clauses were not airtight, cumulatively they constituted a warning that it would be difficult for nonparties to profit from international trade in the controlled substances.

Low-Consuming Developing Countries

Since CFC technology is relatively easy to obtain and install, developing countries, with their rapidly growing populations, represented a large potential source of future CFC emissions. Existing per capita consumption of CFCs in developing countries was only a small fraction of that of the industrialized world, but their domestic requirements were growing. The negotiators at Montreal thus faced a difficult challenge in designing special provisions to encourage developing countries to sign the protocol. The drafters of the treaty needed to enable these nations to meet legitimate needs during a transition period while substitutes were being developed; at the same time, it was important to diminish incentives for them to become major new CFC producers and consumers.

Although this potentially divisive subject had been debated throughout the negotiations, it was not until Montreal, when considerably more developing countries were in attendance, that specific details were considered. Egyptian Ambassador Essam-El-Din Hawas ably chaired a working group to draft a package of relevant provisions; Argentina, Brazil, Kenya, Mexico, Venezuela, and, for the first time, China and Malaysia played leading roles in this effort.

The resultant article 5 postponed but did not eliminate compliance by developing countries. Early proposals to allow a 5-year grace period and an annual per capita consumption cap of 0.1 kilogram were rejected by developing countries as too restrictive. The final compromise permitted a 10-year period during which any developing country with a per capita annual consumption below 0.3 kilogram could increase consumption up to this level "in order to meet its basic domestic needs." This quantity represented approximately 25 to 30 percent of the existing per capita consumption in Europe and the United States, and about 50 to 60 percent of the targeted level in the industrialized countries after their cutbacks were effected during the same period.[18] After the 10 years had elapsed, a developing country would be required to adhere to the article 2 reduction schedule.

The 10 to 15 percent differentials permitted between the scheduled phasedowns of production and consumption for the industrialized countries would theoretically provide producers with the excess capacity to enable them to satisfy developing nations' needs. The protocol also encouraged financial and technical assistance to developing countries for alternative substances and new technologies (article 5, paragraphs 2 and 3; article 9, paragraph 1; article 10), but the vagueness of these clauses was subsequently to prove troublesome.

The negotiators believed that developing countries were unlikely to expand their use of CFCs to the maximum permitted level, since they would not find it attractive to invest in a technology that was both environmentally detrimental and soon to be obsolete as substitutes were introduced. Some analysts, however, were concerned that developing countries might use the permissible increase in consumption of bulk CFCs not merely to satisfy legitimate internal demand ("basic domestic needs") but also to manufacture products containing CFCs for export. Thus, for example, a company in an industrialized country could circumvent the spirit of the protocol by exporting or producing bulk CFCs in a developing country, manufacturing products using those CFCs, and then importing the products or selling them on the world market. Although the protocol did prohibit developing nations, beginning in 1993, from exporting bulk substances to nonparties (article 4, paragraph 2), their negotiators at Montreal had resisted proposals to ban the export of products containing CFCs. The key to this issue was that the concept of "basic domestic needs" was not precisely defined in the protocol and therefore still open to interpretation.

As it turned out, North-South issues were far from settled at Montreal, and they became a central focus of subsequent deliberations over the protocol's implementation.

Special Treatment for the European Community

Substantive differences between the European Community and other governments were complicated by the fact that the EC demanded special concessions for the Commission and the 12 member nations by virtue of their evolving political and economic union. Of central relevance to these demands was the question of how authority on any given matter was divided between the EC Commission and its sovereign member states, a situation that to outsiders seemed perpetually in flux. It was often unclear to other participants in the negotiations, nor could the EC representatives themselves satisfactorily explain, whether the EC Commission had full authority ("exclusive competence") to enforce any given article of the protocol or whether power was shared with member countries ("mixed competence").[19]

As early as 1984, EC Commission legal experts, in consultations with U.S. State Department lawyers, maintained, incorrectly as it turned out, that the proposed control protocol to the Vienna Convention would fall within the exclusive competence of the Commission.[20] In 1986, however, Community Legal Adviser John Temple Lang observed that "precisely because the limits of exclusive competence are politically important, they are particularly difficult and controversial to define." In fact, he added, "vagueness was a natural result of the wish to avoid difficult internal discussions about the extent of Community competence."[21] Other European scholars pointed out that even the 1987 Single European Act "did not give the Community exclusive competence in this field [the environment]."[22] Adding to the confusion was the fact that, if a protocol was found to be within the Community's exclusive competence, "no member state of the Community could be permitted (under Community law) to ratify it."[23]

The EC Commission's insistence on special statutory treatment as a "regional economic integration organization" (REIO) thus became an added irritant to its negotiating partners. Even while other governments applauded the philosophy of European union, they could not in the case at hand be certain whether the ambiguities might, under a yet unforeseen circumstance, allow both the Commission and an EC member state to maintain that a given treaty obligation was not its responsibility. Under these circumstances, even Temple Lang conceded that "a lawyer from a non–Member State is likely to have great difficulty in using the Community's declaration of competence to answer the question whether the Community's ratification will or will not lead to its being bound by any given article"; he further acknowledged that "non–Member States may be legitimately interested in knowing which obligations are undertaken, at a given moment, by the Community and any of its members."[24]

During the negotiations leading up to the Vienna Convention, it seemed to many observers as if the EC Commission delegation was concentrating, not on the substance of ozone layer protection, but rather on legal arguments related to the terms by which the Community could become a party to the convention.[25] The REIO issue had, in fact, almost scuttled the 1985 Vienna Convention,[26] and the EC Commission raised it again, in a new form, late in the protocol negotiations. The EC representative at the July 1987 legal experts' meeting in The Hague insisted that the Community be treated as a single unit for purposes of determining compliance with all control obligations, including the right to submit all required data on production and consumption in aggregated form for the 12 member countries. The rationale offered was that, after the full establishment of a European single market in 1992, trade among individual EC member states could be neither restricted nor even reported. However, when asked by other delegations to furnish precedents from relevant international agreements for their proposal for single-unit treatment, EC lawyers were unable to do so.[27] Nor could they definitively state that the central executive of the European Community had the power to enforce compliance of its sovereign member states with environmental treaty provisions.

If the EC were to join the protocol as a single unit, with full power to enforce all treaty obligations within its borders and with no separate participation in the protocol by individual EC member states, it would correspond in status to the U.S. federal system. But the EC Commission had no intention of excluding its member states from the protocol. On the contrary, it insisted that each of the 12 member governments be voting parties to the protocol.

In the closing hours of the Montreal negotiations, other delegates (including some from EC member governments) were exasperated that, after all the difficult substantive differences had finally been reconciled, the Commission representative seemed prepared to block the treaty unless further concessions were granted. The United States, along with Australia, Canada, New Zealand, and others, resisted the Commission's proposal. They observed that, if the EC were treated as a unit for purposes of compliance, reductions already planned by West Germany could be offset by unchanged or even expanded output by other EC members known to be reluctant to impose controls. The Austrian chairman of the Montreal conference, Winfried Lang, himself a legal expert, agreed that this was a possible consequence: he wrote in 1988 that "the aggregated accounting for EC consumption enables individual EC states to lag behind in their [treaty] obligations to the extent that other EC states undertake their reductions more rapidly than required by the Protocol."[28]

The EC proposal could thus have allowed those state-parties that happened to be members of a REIO ostensibly to undertake a protocol responsibility but then in practice to avoid it. Further, if the EC insisted on submitting aggregated rather than country-by-country data, it would be impossible for other parties to monitor the compliance of individual EC member states. The United States declared that it would not acquiesce in the creation of a legal system that could permit some parties to enjoy the political benefits of adhering to a treaty without being unequivocally subject to the treaty's discipline.

After a nerve-racking midnight standoff over this issue, during which the fate of the protocol hung in the balance, a compromise was achieved at the last possible moment. New Zealand's environment minister, Philip

Woollaston, suggested that the protocol allow EC members to be treated as a unit for purposes of consumption—but not for purposes of production (article 2, paragraph 8). Moreover, this concession would obtain only if all member countries plus the EC Commission became parties to the protocol and formally notified the secretariat of their manner of implementation. Woollaston's formulation neatly avoided the EC's alleged difficulty in coming up with intra-European export and import figures after the 1992 single market was achieved. Since export and import data were required only for calculating a country's allowed consumption level (production plus imports minus exports), individual EC countries need not be relieved of the obligation to report separately on production.

During the discussions leading up to the REIO compromise, other delegations pointed out an inconsistency between the EC demands and the fact that the Commission could not even guarantee that all EC member states would in fact sign, or subsequently ratify, the protocol. Indeed, only nine member governments were actually participating in the Montreal negotiations. With heroic efforts, the EC delegation managed in the final hours to round up missing representatives of Greece, Ireland, and Portugal—local consular officials and a third secretary of embassy from Ottawa—but of these only Portugal received official authorization to sign the protocol, and Ireland was even erroneously omitted from the official UN attendance list appearing in the Montreal Protocol "Final Act." As Luxembourg also did not sign, the Commission was able in the end to sign the protocol on September 16 with only nine of its member states.

Variations of the REIO theme were the final matters to be resolved both in Vienna and in Montreal, and on both occasions an impasse was only narrowly averted. Despite the compromise at Montreal, REIO issues were to emerge again later (see Chapters 9 and 10). The principles involved clearly go beyond environmental protection, and REIO problems are likely to continue to challenge and perplex treaty negotiators unless mutually acceptable formulas are devised.

The Immediate Aftermath

8

On September 16, 1987, representatives of 24 nations, plus the Commission of the European Communities, signed the Montreal Protocol on Substances That Deplete the Ozone Layer. Although most were industrialized countries, the signatories also included Egypt, Ghana, Kenya, Mexico, Panama, Senegal, Togo, and Venezuela. In the following months, many other governments added their signatures. Signing a treaty, however, is only the first step—a declaration of intent. The proof would be in the formal ratifications still to come, for until a state actually ratifies a protocol, no binding commitment exists under international law.

Ambiguity and Flexibility

The road to Montreal had been both difficult and instructive, and many obstacles had been overcome. Ironically, if a control protocol had been agreed upon at the Vienna conference, two and a half years earlier, it would not have been as strong. Although the Montreal Protocol was not perfect, it did represent a new approach to dealing with environmental risks.

There were, to be sure, ambiguities and weaknesses in the final text.[1] For example, there remained open a theoretical possibility that chemical companies might transfer production to low-consuming countries or even to nonparties. In addition, the three-to-five-year intervals before the parties had to agree on definitions determining restrictions on imports of products containing or made with chlorofluorocarbons could somewhat weaken the pressures on nonparties to join the protocol. The monitoring and compliance clause (article 8) also needed further elaboration, although it is noteworthy that this provision was itself unusual for a treaty outside the realm of arms control. Key terms of the protocol were also left undefined at Montreal, including "developing countries" and "basic domestic needs."

These and similar issues remained to be worked out at future meetings of parties to the protocol. The negotiators at Montreal had made their top priority the setting into motion of an international process to reduce emissions of ozone-depleting chemicals. Therefore, they had consciously decided not to try to resolve every conceivable contingency, because to do so would have involved further delay. Rather, they moved to establish the protocol firmly in international law as rapidly as possible.

A unique strength of the treaty was, in fact, that it was deliberately designed to be reopened and adjusted as needed, on the basis of the periodically scheduled scientific, economic, environmental, and technological assessments. This flexibility—a streamlined process for modifying the protocol—was a major innovation (article 2, paragraphs 9 and 10, and article 6). The timing and extent of reductions, as well as the chemicals covered, were primed for change as scientific and other factors evolved. These provisions not only reflected the scientific uncertainties at the time of the negotiations but also served to bridge differences between those governments demanding strong controls and those who were less convinced. The latter could point out to critical constituencies (for example, industry) that the protocol's flexibility did not operate in one direction only: controls could also be relaxed if the unfolding science were to indicate a diminished threat to the ozone layer. In a real sense, the Montreal Protocol was, as foreshadowed during brainstorming sessions at the 1986 Leesburg workshop, an "interim protocol."

Second Thoughts from the Third World

Because of the global implications of local CFC and halon production, wide participation in the treaty would be essential to its success. The major CFC-producing and -consuming countries had recognized that their actions alone would be insufficient to protect the ozone layer. Therefore, the protocol was designed with incentives to encourage as many developing nations as possible to accept its objectives. From the interest shown by developing-country delegates at Montreal, including even those who had not received authorization to sign on the spot, it seemed likely that most would eventually ratify.

Both during and after the negotiations, a number of developing nations were at the forefront in demonstrating a commitment to protect the ozone layer.[2] Mexico became the first nation to ratify the protocol, and it moved quickly to reduce CFC and halon imports and production. Venezuela convened a conference of Latin American countries to promote implementation of the treaty. Egypt announced that its only CFC-using factory would switch to other products. In Indonesia, the environment minister blocked an application for what would have been the country's first CFC facility. Kenya and Thailand announced plans to limit imports.

However, not all developing countries were satisfied with the protocol. The Malaysian negotiator at Montreal, Abu Bakar Jaafar, characterized the treaty as "inequitable." He wrote that developing nations "had been had" at Montreal because they "were unaware of the full socioeconomic implications of the protocol." He stated that there had been insufficient time to review the "lengthy and legalistic English-language protocol" before it had been "rushed to the plenary session for signature." Jaafar criticized the protocol for allotting a lower per-capita consumption quota to developing countries than to industrialized countries. He also contended that the treaty's trade provisions amounted to "trade war by environmental decree."[3] Although Malaysia did ratify the protocol in August 1989, its objections influenced other governments.

In a mirror image of complaints by some developing countries that the protocol was unfair to them, other observers criticized the treaty for making overly generous concessions. To demonstrate their point, these analysts calculated "worst case" scenarios of ozone depletion if all developing countries were to reach their allowable per capita CFC and halon use—an assumption that seemed at the time rather dubious.[4]

In this connection, however, a worrisome factor was that the accession of the two most populous nations, China and India, remained in doubt. Although these two countries were minimal producers and consumers of CFCs, together they accounted for nearly 40 percent of the world's population. Accordingly, they could undermine the treaty were they significantly to expand their use of CFCs for refrigeration and other purposes. The protocol's restrictions on international trade in these chemicals would have no inhibiting effect on China and India because of their huge potential domestic markets.

China had made substantive contributions to the deliberations in Montreal and had conveyed some positive signals at the ministerial level concerning future accession. More troubling was the possible stance of India, which had shown no interest in the negotiations and whose officials in private conversations had characterized the issue as a "rich man's problem—rich man's solution." India had bought CFC technology from the U.S. firms Allied Chemical and Pennwalt before the Montreal Protocol negotiations began. By 1989 there were at least four factories in India, two of which were built after the protocol was signed. Since domestic demand did not justify the availability capacity, Indian entrepreneurs were seeking foreign markets and reportedly had sold technology to Iraq.[5]

Although nonaccession to the protocol did not necessarily ensure that a country would become a pollution haven, this possibility was clearly a matter of concern. The issues of Chinese and Indian accession and of equitable treatment of developing nations were destined to loom larger as the sense of urgency over the ozone layer deepened and the parties began to face the realities of implementing the protocol.

Ozone *Glasnost*

One of the most interesting and overlooked aspects of the ozone history was the manifestation during the negotiations of the Soviet Union's new policy of *glasnost.* The Soviet government's interest in the environment had been evolving, as demonstrated by its willingness after the Chernobyl disaster to reveal data and to cooperate in the 1986 International Atomic Energy Agency review of the accident. Also during this period, consultations and technical exchanges under the U.S.-Soviet bilateral environmental agreement were revived.

At the opening of the protocol discussions in December 1986, the U.S. delegation had sensed that the Soviets were unfamiliar with the latest science on the potential dangers of CFCs. Therefore, taking the occasion of an informal lunch two weeks later with a visiting Soviet minister at Washington's Cosmos Club (appropriately, an institution founded in the nineteenth century to promote scientific endeavors), U.S. Assistant Secretary of State John Negroponte and the U.S. chief negotiator proposed a collaborative research effort on ozone and climate. The resultant meetings of Soviet and U.S. scientists in the spring of 1987 contributed to a gradual weakening of Soviet opposition to international controls on ozone-depleting compounds. Throughout the ensuing protocol negotiations, relations between the Soviet and American delegations were unusually cordial and marked by frequent informal consultations. (One such exchange at Montreal led to the resolution, described in Chapter 7, of the base-year problem that had threatened to stall the negotiations.)

After the signing at Montreal, expanded bilateral cooperation in ozone and climate research was reflected in the Reagan-Gorbachev summit discussions and communiqués in both 1987 and 1988.[6] This scientific collaboration both accompanied and reinforced the gradual political rapprochement between the two governments. Previously off-limits cloisters of the U.S. National Aeronautics and Space Administration and the Soviet Space Research Institute became host to mutually beneficial meetings between Soviet and American colleagues. Indeed, there were unexpected dividends. As the U.S. Nimbus-7 satellite—which carried crucial ozone-measuring instrumentation—approached the end of its space life, NASA found that it would be impossible, because of budget constraints and lack of launch vehicles, to get a replacement in orbit before the Nimbus-7 expired. The Soviets offered to help by carrying American ozone-measuring equipment aloft in 1991 in a Russian weather satellite.[7]

Early Public and Industry Responses

Initial reactions to the Montreal Protocol from the media and citizens' groups were highly favorable and recognized its precedent-setting nature. A *Newsweek* headline labeled it "An Exemplary Ozone Agreement," while the *Washington Post* called the treaty "an extraordinary achievement" and uncharacteristically gave the Reagan administration "enormous credit" for its leadership on the issue.[8]

Most environmental groups also praised the protocol. Typical of the reaction was the statement of an atmospheric scientist at the Environmental Defense Fund that "this is as important as an arms agreement."[9] A few environmentalists in the United States did complain that the protocol could have been stronger; some were critical of the "loopholes" for the Soviet Union and for developing countries.[10] Within weeks of the treaty's signing, the city council of Berkeley, California, banned fast-food foam packaging made with CFCs, an action that, followed by similar measures in many other localities, gave American industry a clear indication of the public mood.[11]

Although European chemical industry observers at Montreal appeared shaken when it finally became clear that their long battle to block an effective international agreement had been lost, U.S. industry seemed relieved that deadlock had been avoided and a reasonably level playing field secured. The Alliance for Responsible CFC Policy termed the accord "an unprecedented step to protect the global environment."[12]

The CFC Alliance continued to assert, however, that "current use of the compounds presents no significant risks to health or the environment" and that the protocol therefore provided a substantial margin of safety. Indeed, the CFC Alliance expressed concern that the treaty's reduction schedule attempted "to go too far, too fast, and far beyond that which is necessary based on current scientific understanding." The industry association initially predicted that costs of implementing the protocol in the United States would be in the range of $5–10 billion; later estimates by other industry sources ranged much higher, although all were necessarily in the realm of guesswork.[13] Even a respected analyst from the environmental community observed in 1988 that "those who argue that the Montreal Protocol was possible because finding substitutes for CFCs is 'easy' arguably know very little about the technical issues still to be addressed."[14]

Some industrialists were tempted to seek ways of evading the protocol's intent before it became international law. There were reports that some companies were selling CFC technology to developing countries. French commentators observed that the French firm Atochem, rather than expanding research into CFC substitutes, "seems to prefer to deploy most of its efforts toward maximum use of exceptions provided for by the Protocol." The French researchers reported that Atochem, which was the largest single CFC manufacturer in Europe, had entered into negotiations with West German, Greek, and Tunisian interests with the aim of transferring CFC production to nonparties or to developing countries exempt from the cutbacks.[15]

A public relations brochure issued by a major British chemical company a few months after Montreal claimed that current CFC emissions posed "no risk" to the ozone layer, cast doubt on ill effects from ozone layer depletion, and warned of job losses and unspecified dangers to consumers from CFC substitutes if there were "any rash further restrictions . . . beyond the prudent measures embodied in the United Nations agreement."[16] British industry, however, was roundly criticized in the House of Lords for continuing to use in its publicity "hackneyed and discredited" language in describing CFCs as being "thought to damage" or "alleged to damage" the ozone layer.[17]

After the protocol's signing, a U.S.-based environmental group, Friends of the Earth, launched a campaign in the United Kingdom calling for consumer boycotts against aerosol sprays containing CFCs. In response, two CFC-producing companies, Imperial Chemical Industries and ISC Chemicals, joined with the British Aerosol Manufacturers Association in a public complaint about anti-CFC newspaper advertisements. The industry continued to insist that "there is no evidence of a link between CFCs and cancer."[18]

As late as June 1988, U.K. government officials argued in Parliament against mandatory labeling of spray cans. Only after Johnson Wax—a U.S. subsidiary in the British aerosol market—announced that it would label its spray cans as not harmful to the ozone layer did the U.K. aerosol industry change its policy. Responding to the new competitive situation, the local producers yielded to mounting consumer pressures and began phasing out CFCs as spray propellants.[19]

It was evident, however, that the protocol was in fact moving industry in directions that two years earlier had been considered impossible. Both producer and user companies began racing to find substitutes for CFCs, and newspaper accounts of new research efforts and promising leads began to appear with regularity.[20] Four months after the Montreal conference, several hundred industry representatives gathered at a trade fair in Washington cosponsored by the U.S. Environmental Protection Agency, The Conservation Foundation (a private American policy research instiute), and Environment Canada. The meeting attracted an overflow audience to lectures, exhibits, and demonstrations aimed at stimulating research and exchanging information on alternatives to CFCs.

AT&T, in cooperation with a small Florida company, announced a substitute for CFC 113, derived from citrus fruit, that could clean electronic circuit boards. Du Pont unveiled plans to construct a pilot plant for a new generation of refrigerants and announced that it would build a facility in the United Kingdom to provide European markets with the CFC-free aerosols that had been standard in the United States for a decade. At the initiative of U.S. companies, an international consortium was established in late 1987 for toxicity testing of CFC substitutes; the multimillion-dollar cooperative program would soon include 16 chemical companies from the United States, Europe, Japan, and South Korea.[21] Similarly, major halon producers from France, West Germany, the United Kingdom, and the United States joined with fire equipment industry associations in a new research institute for joint testing of alternatives.

The Montreal accord was clearly functioning as its designers had intended; none of these actions would have occurred in the absence of the protocol. Indeed, industry's post-Montreal response vindicated the U.S. decision late in the negotiations not to risk delaying agreement on international controls by continuing to insist on a firmly scheduled CFC phaseout. Even most U.S. allies had considered this step unnecessary at the time, and gaining the concurrence of the European Community, the Soviets, and many others would have been impossible.

By cutting the market in half at a fixed date, the protocol was in fact tipping CFCs toward obsolescence. U.S. negotiators had reasoned that, when substitutes were developed to such an extent, the remaining CFC market could probably not be sustained. By providing CFC producers with the certainty that their sales were destined to decline, the protocol unleashed the creative energies and considerable resources of the private sector in a search for alternatives. The treaty at one stroke changed the market rules and thereby made research into substitutes economically worthwhile. Stimulated by market incentives, industrial firms began deploying their resources to find solutions rather than to obstruct regulation. Eventually, industry's plaints regarding the costs and difficulties of adapting to the new controls turned out to have been vastly overstated.

New Debates in the European Community

Following the signing of the protocol, the U.S. government sustained its rapid pace of action. Within two months, EPA had issued proposed regulations on CFCs and halons for public comment, and the Senate approved the treaty by a resounding 83–0 vote in March 1988.

Inside the EC, however, it became apparent that even after Montreal, differences over ozone persisted. At a critical meeting in December 1987, the Federal Republic of Germany, together with Denmark and the Netherlands, urged the Community to move beyond the Montreal Protocol, with an ultimate objective of eliminating CFCs.[22] Other countries, however, argued for a "more restrained Commission position, aimed only at fulfilling the obligations of the Protocol." A contemporaneous Dutch government publication complained that "the impression exists that most EC member states do not have a clear policy for reducing CFC use; the initiative is being left to industry and the European Commission."[23]

At the December 1987 meeting, the United Kingdom argued that the EC as a unit, rather than each individual country, should meet the protocol's requirements for production cuts. This proposition disregarded the struggle at Montreal that had resulted in the compromise specifically allowing Community members to meet treaty obligations jointly *only* with respect to consumption. Although the British disingenuously pointed out that the protocol "text is silent on joint fulfillment of production," their proposal failed.[24] Echoing the concerns expressed by non-EC delegations in Montreal, Denmark, the Federal Republic of Germany, and the Netherlands insisted that an EC "bubble" would be unacceptable—that is, no individual EC member could be permitted lower standards than required by the treaty, whether for consumption or production, even if they were balanced by more stringent reductions elsewhere in the Community.

Europe Discovers Nonessential Aerosols

A curious aftermath to Montreal was the attempt by some European commentators and officials to portray the ozone treaty as a triumph for the European Community's original approach. One British observer contended that the EC could "claim the credit for being the architect of a central pillar of the Protocol, namely the production limit"; the EC had "won the intellectual argument," as the Americans had been persuaded to abandon their position on a global aerosol ban and to accept the logic of the 1980 EC policy of setting a limit on production.[25]

This thesis overlooked the essence of the conflicting U.S. and EC positions in 1985 (see Chapter 4). The United States had never opposed controlling production per se; rather, it had recognized that the EC concept would have permitted unrestrained output by the Community for another two decades. The Toronto Group had favored an immediate global ban on aerosols in 1985 simply because it was economically and technically feasible and would have provided substantial short-term protection for the ozone layer. Such a step would also have bought time for further scientific research to establish a basis for international agreement on more stringent overall production controls. In fact, had the EC followed the U.S. lead in banning CFC aerosol sprays in 1978, the world's CFC output during the period 1979–1987 would have been approximately 30 percent, or 2 million metric tons, lower.[26] Such a reduction obviously would have benefited the ozone layer enormously.

As for EC leadership on the ozone issue, in 1988 the U.K. House of Lords "recognise[d] that countries outside the Community took the initiative in pressing the Community to make significant reductions."[27] A European scholar was less diplomatic, characterizing the EC attitude toward ozone protection as "very inflexible and reluctant during the whole debate preceding the signature of the Montreal Protocol." He concluded that, "after long and staggering negotiations, strong U.S. pressure had finally forced the EC to accept an agreement."[28]

Ironically, the U.S. government, in line with its perception of mounting danger to the ozone layer, had urged deep production cuts during the protocol negotiations in full knowledge that it would be much harder for American than for European companies to implement such restrictions. Since aerosols still accounted for over 50 percent of European CFC production, EC companies could take the same steps that Americans had taken 10 years earlier—and thereby easily attain the Montreal Protocol's 50 percent reduction goal far in advance of the 1998–99 treaty timetable. In contrast, U.S. cuts would have to come in more difficult areas such as air conditioning and refrigeration.

Suddenly the European chemical companies and user industries, which for over a decade had been informing their governments that CFCs were indispensable in spray cans, discovered—under unaccustomed pressure, to be sure, from aroused consumers and legislators—that the U.S. premise on an aerosol ban made sense after all.[29] By 1988, companies in Belgium, the Federal Republic of Germany, the Netherlands, and the United Kingdom voluntarily agreed to begin phasing out CFCs as propellants. And the U.K. House of Lords recommended a mandatory Community-wide ban on CFCs in aerosols as "the quickest and easiest way to produce a significant reduction."[30]

An official statement on aerosols presented after Montreal by a European industry association to the EC Commission said it best: "Phasing out uses of CFCs in applications where proven safe alternatives exist is both feasible and desirable." An aerosol ban could be achieved with "minimal anti-competitive or socio-economic disruption to the Community," since "European industry is in the fortunate position of being able to introduce the alternative techniques developed [in America] without suffering the same degree of hardship."[31]

And U.S. industry, in order to meet the new protocol's targets, was thrust into the position of having to develop substitutes for CFCs in uses for which no alternatives currently existed.

New Science, New Urgency

9

Even as negotiators were hammering out the final compromises in Montreal, an unprecedented international scientific expedition was taking place in and over Antarctica. Spearheaded by the National Aeronautics and Space Administration, the National Oceanic and Atmospheric Administration, and the National Science Foundation, and also financed in part by the U.S. Chemical Manufacturers Association, some 60 scientists of different nations spent several weeks during August and September 1987 in the subfreezing and stormy conditions of the austral early spring. Based in Punta Arenas, Chile, the scientists employed specially designed equipment, placed in balloons, satellites, a DC-8 flying laboratory, and a unique converted high-altitude U-2 aircraft, to track stratospheric chemical reactions and minute concentrations of gases over Antarctica. The results of this expedition would simultaneously vindicate the efforts to achieve the control protocol and reveal it as inadequate, in its existing form, to protect the ozone layer.

At a press conference on September 30, 1987, just two weeks after the signing of the Montreal Protocol, Robert Watson of NASA and Daniel Albritton of NOAA released interim findings from the Antarctic expedition. The data revealed an apparently substantial worsening of the seasonal ozone depletion that had been earlier measured by British scientists from ground-based observations (see Chapter 2), as well as an abnormal presence of chlorine. However, this initial report stopped short of definitively attributing the ozone hole solely to CFCs. It noted that the role of natural atmospheric circulation was still not fully understood, that other theories had not yet been invalidated, and that further intensive analyses of the preliminary data would be undertaken.[1]

Setting the Stage

In January 1988 Mostafa Tolba convened a meeting in Paris of about a dozen senior advisers from governments, environmental organizations, and industry to consider practical details of implementing the Montreal Protocol.[2] This group determined that the overriding immediate issues were the early entry into force of the treaty and the broadest possible participation of countries. At this time, no nation had yet ratified the protocol. It was by no means inevitable that the requisite number of major consuming countries would ratify during 1988 so that the treaty could enter into force by the target date of January 1, 1989. Tolba agreed to use his personal influence with governments to achieve that objective.

The group also identified textual ambiguities and data needs that required early attention from governments, and it created an implementation timetable. First would come a meeting of experts in Nairobi in March 1988 to determine procedures for governments to report production and consumption data on chlorofluorocarbons and halons to the United Nations Environment Programme; this was essential for establishing the 1986 base-year data, which were crucial both for the protocol's entry into force and for the application of control measures. Three important meetings were then scheduled for October 1988 in The Hague: a scientific symposium to update findings on the state of the ozone layer, a meeting of legal and technical experts to develop recommendations for the First Meeting of Parties to the protocol, and a technical workshop for industry on the status of research into substitutes and alternative technologies.

Significantly, the group recommended advancing two important dates. The First Meeting of Parties was rescheduled from November to May 1989. And the critically important scientific, environmental, economic, and technological assessments would be completed not in 1990, as originally scheduled, but in 1989, in order to provide governments with comprehensive analyses sooner than had been foreseen only four months before by the signatories at Montreal. The stage was being set for the earliest possible revision and strengthening of the Montreal Protocol.

Thus, even before the treaty formally entered into force, UNEP was continuing its role of keeping governments focused on the evolving environmental danger. This advance activity proved to be both farsighted and fortunate. Scientific understanding of the ozone layer continued to progress—and in a fashion that dramatically justified the heretofore primarily theoretical concerns that had motivated the drive for international controls.

Alarming New Evidence

Just six months after the provisional results of the Antarctic expedition had been announced, another joint NASA-NOAA press conference sealed the fate of CFCs and halons. On March 15, 1988, the report of the Ozone Trends Panel was released—a 16-month comprehensive scientific exercise involving more than 100 scientists from 10 countries, using new methods to analyze and recompute all previous air- and ground-based atmospheric trace gas measurements, including those from the recent Antarctic expedition.[3] The panel's conclusions made headlines around the world. Ozone layer depletion was no longer a theory; at last it had been substantiated by hard evidence. And CFCs and halons were now implicated beyond reasonable doubt.

Reanalysis of the data demonstrated conclusively that human activities were causing atmospheric concentrations of chlorine to increase on a global scale. Even more disturbing was the finding that, from 1969 to 1986, a small but significant depletion of the ozone layer, amounting to 1.7 to 3.0 percent (depending upon latitude), had already occurred over heavily populated regions of the Northern Hemisphere, including North America, Europe, the Soviet Union, China, and Japan. The decline ranged from 2.3 to 6.2 percent (depending upon latitude) during winter months.

With respect to Antarctica, the Ozone Trends Panel confirmed that a "large, sudden, and unexpected" decrease in springtime ozone had developed over the previous decade. Precipitous localized losses of up to 95 percent had been observed during the 1987 expedition. Further, the total column of ozone measured in the spring of 1987 had been the lowest since observations had begun 30 years earlier, and the low ozone phenomenon had persisted longer than ever recorded. The panel concluded that the evidence "strongly indicates that man-made chlorine species are primarily responsible for the observed decrease in ozone."[4]

For the first time, scientists understood how chlorine released from CFCs could trigger the springtime collapse of the Antarctic ozone shield. This hitherto elusive mechanism involves "heterogeneous chemistry"—

unusually rapid chemical reactions involving chlorine that occur on ice crystal surfaces in polar stratospheric clouds. These chemical processes are activated by the return of sunlight and ultraviolet radiation following the extremely cold night of the Antarctic winter. They continue until the growing warmth of the sun finally breaks up the previously isolated vortex of polar winds within which the ozone-destroying process takes place.

The Antarctic observations aroused suspicions that similar, if less pronounced, chlorine reactions could also occur over the Arctic. Plans were announced for an international scientific mission in the winter of 1988–89 to test this hypothesis.

The implications of the new scientific findings were profoundly disquieting. The model projections underlying the control provisions of the Montreal Protocol had assumed a probable global average ozone loss of around 2 percent by the middle of the twenty-first century. Now it was revealed that more than this had already occurred, and, indeed, that ozone depletion appeared to be accelerating with increased accumulation of atmospheric chlorine. The existing models had proved incapable of predicting either the chlorine-induced Antarctic phenomenon or the extent of ozone depletion elsewhere over the planet. They were therefore probably *underestimating* future ozone losses.[5]

Thus, the state of the science had fundamentally changed. A sense of uncertainty about the models' reliability made the future of the ozone layer seem even more precarious. The one new certitude was that further significant depletion of the ozone layer would occur even if the Montreal Protocol were fully implemented by all nations.

Calls for Phaseout

As a result of the Ozone Trends Panel report, pressure for a phaseout of CFCs began to mount. Perhaps the most unexpected early response came from Du Pont, the world's largest producer. Only 11 days before the NASA-NOAA March 1988 press conference, Du Pont's chairman, replying to a letter from three U.S. senators reminding the company of its 1974 pledge (see Chapter 2) and suggesting that it halt CFC production, had declared, "At the moment, scientific evidence does not point to the need for dramatic CFC emission reductions."[6] Less than three weeks later, however, citing the new scientific findings, Du Pont announced that it would substantially accelerate research into substitutes and would stop manufacturing all CFCs and halons by the end of the century. Before either the European Community or the U.S. government, Du Pont proposed an international phaseout of the chemicals.[7] Six months later, in September 1988, U.S. EPA Administrator Lee Thomas, citing a new study by his agency, called for complete global elimination of both CFCs and halons, plus a freeze in use of methyl chloroform, a popular industrial solvent not previously recognized as a major source of stratospheric chlorine.[8]

Appeals to ban CFCs were further stimulated by heightened public concern in the summer of 1988 over the prospect of rapid climate warming. Although scientists had known earlier about the heat-trapping effect of CFCs, the climatic anomalies of 1988, which included record heat, drought, and storms, had dramatized to the general public the possible effects of changes in global climate. It was estimated that CFCs contributed as much as 25 percent of the total greenhouse gas effect.[9]

That the CFCs were greenhouse gases was bad enough. But, since carbon dioxide, methane, and other trace gases that partially offset the effect of CFCs on stratospheric ozone are themselves greenhouse gases, any future measures taken to limit their emissions in order to mitigate global warming would actually exacerbate ozone destruction—unless chlorine concentrations were significantly reduced.

Reacting to the new scientific data, a few environmentalists forgot how uncertain the scientific evidence had actually been during the protocol negotiations; they also seemed to misunderstand the treaty's capability to evolve as those uncertainties were narrowed. One U.S. observer termed the protocol "a major half-step forward," while a British writer uncharitably described it as "a masterpiece of fudge and compromise . . . full of loopholes" and "a feeble agreement."[10]

In this atmosphere of public anxiety, some American activists urged immediate renegotiation of the protocol. UNEP rightly resisted panicky appeals to convene an emergency meeting.[11] At a time when the treaty had not even entered into force and very few countries had ratified it, reopening negotiations might only cause other governments to delay ratification and adopt a wait-and-see attitude, or even frighten them away altogether.[12] The first priority remained ratification by as many countries as possible and entry into force on January 1, 1989. The planned process was well under way. The meetings scheduled for The Hague in October 1988 would be even more important in light of the new discoveries. And the protocol itself contained carefully designed procedures for its own revision. Tolba wisely stayed the course, allowing the scientists further to consolidate their analyses and the political pressures for stronger controls gradually to gain worldwide momentum.

The European Community Changes Direction

Within the European Community, the new scientific discoveries had a decisive influence. During the late spring and summer of 1988, environmental groups throughout Europe, led by affiliates of the U.S.-based Friends of the Earth, for the first time made ozone protection a priority issue.

The West German government, again at the forefront, announced a unilateral goal of 95 percent reduction. It further proposed that the Montreal Protocol be revised to attain an 85 percent cut by 1995, with a firm objective of complete elimination of both CFCs and halons. Bonn based its new policy on a comprehensive report by a special multiparty West German parliamentary commission, released in the summer of 1988 after several months of expert testimony and deliberations.[13]

Germany rushed to exploit the power of its EC presidency to achieve a stronger Community-wide position on ozone protection before yielding the office to Greece on July 1, 1988. On June 16, the EC Council formally decided that the Community would ratify both the Vienna Convention and the Montreal Protocol. The Council also approved detailed regulations for implementing the protocol. In a response to the West German appeal, it passed a resolution calling on member states to undertake voluntary agreements with industry for reductions beyond the 50 percent required by the treaty. The Council also stated definitively that there would be no EC "bubble" under which some members could offset CFC increases against reductions by other members. It thus put to rest a long-standing concern to observers both inside and outside the Community.[14]

But not all members of the EC Council had yet come as far as West Germany. The Community refrained at its June meeting from formally endorsing a strengthening of the Montreal accord. The recommendation for voluntary CFC reductions "to the maximum extent possible" was a nonbinding compromise.[15] And the Council continued to resist appeals from Denmark, the Federal Republic of Germany, and the Netherlands for mandatory product labeling.

A surprising development, however, was the galvanizing effect of the new scientific evidence on the U.K. government later that summer. Within a few months, the United Kingdom was transformed from a reluctant follower to a world leader in the drive to protect the ozone layer. It was a remarkable and reassuring development, demonstrating that, with leadership from the top, it is possible for a government to jettison policies supported by powerful economic interests and instead promote long-term environmental protection. The British royal family had actually anticipated the government's move by several months; Prince Charles had announced on television earlier that year that his household would no longer use CFC aerosol sprays.

A major factor in this policy shift was that during the summer of 1988 British scientists finally received the serious political hearing that had long been denied them. The United Kingdom Stratospheric Ozone Review Group, in an analysis based on the Ozone Trends Panel report, independently concluded that the reductions contemplated in the Montreal Protocol would be insufficient to prevent further significant depletion of the ozone layer. The British scientists strongly recommended phasing out both CFCs and halons.[16] These findings had a profound influence on British leaders of opinion.

Stimulated by pressure from British environmental groups, especially national affiliates of Greenpeace and Friends of the Earth, important parliamentary hearings were held in early summer 1988. Both houses of Parliament criticized the government's long-standing position and urged that the United Kingdom press the EC to endorse an 85 percent reduction.[17] It had become clear to many Britons that, merely by phasing out CFC aerosols (which accounted for over 60 percent of British production), the United Kingdom could reach the protocol's 50 percent reduction target far ahead of the 1998–99 protocol schedule.[18]

In September 1988 Prime Minister Margaret Thatcher delivered a pivotal address to the Royal Society, Britain's premier academy of sciences. She emphasized that the U.K. government would henceforth accord higher priority to environmental protection and, in particular, to threats to the ozone layer and global climate. Further, she announced her intention to host an international meeting at the ministerial level in March 1989, aimed at encouraging more countries to ratify the Montreal Protocol and at developing political support for strengthening its terms.[19]

At the EC Council meeting of November 1988, the British joined the environmental progressives, calling on the Community to accelerate the Montreal timetable and to enact an 85 percent reduction as soon as possible. European unanimity, however, still proved unattainable. A British minister later publicly complained that "one or two [member states] have still to be convinced of the overwhelming scientific evidence."[20] France and Spain were reportedly the remaining holdouts.[21] One European analyst suggested that the split between the United Kingdom and France occurred at this point because ICI "had lifted its total opposition to further reductions" while Atochem had not.[22]

Racing the Ratification Clock

While all of this was happening, the Montreal Protocol had still not become binding international law. The UNEP-sponsored meetings in The Hague in October 1988 served further to focus the attention of governments and world public opinion on the unfolding situation. Substantial progress was achieved in clarifying issues that needed resolution by the parties to the protocol soon after it would enter into force. The Hague meetings also established terms of reference and timetables for four separate panels that would assess the evolving scientific, economic, environmental, and technological data. These groups, which were eventually to involve hundreds of experts from many countries, would be developing the analytical bases, as foreseen by the treaty, for the now-anticipated recommendations for revising and strengthening the protocol. The process was to be truly international, designed to develop fresh consensus in an exercise that seemed continually to break new ground in intergovernmental relations.

Unfortunately, however, ratification of the Montreal Protocol proceeded only slowly during 1988. As noted in Chapter 7, the treaty could enter into force only if at least 11 countries, representing at least two-thirds of the world's CFC consumption, formally ratified it, and January 1, 1989, had been established as the target date. A quirk in the protocol specified that, if the requisite conditions were not met by that date, the protocol's entry into force would be postponed until 90 days after such conditions were fulfilled (article 16). Thus, missing the deadline by even one day would delay entry into force until at least April 1989.

In March 1988, Mexico became the first country to ratify, followed three weeks later by the United States. At midyear, only three other countries—Norway, Sweden, and Canada, all relatively small consumers—had joined them. By the end of November, 16 countries had ratified. Belying apprehensions of some environmentalists, Japan and the Soviet Union were among the early ratifiers. In all, however, the ratifying governments still accounted for only about half of global CFC consumption. With one month to go, the large European nations—France, the Federal Republic of Germany, Italy, and the United Kingdom—were conspicuously missing, and without at least two of them joining, the protocol could not enter into force.

The reason for this situation was that the European Commission had persuaded the Council of Ministers to accept an unusual proposal that all member states, plus the Commission, ratify the Montreal Protocol *simultaneously.* The Commission maintained that simultaneous ratification was essential to avoid a putative conflict between EC single market policy, which prohibited internal trade barriers, and the protocol's requirements to restrict trade with nonparties.

This argument was specious. A theoretical inconsistency could exist if some EC members became parties to the treaty while others were still in process. But this situation could plausibly have been resolved by a notification to the protocol secretariat, since even nonratifying EC member states would actually be in compliance with treaty provisions because of internal Community regulations. A more likely explanation for the insistence on simultaneous ratification was the Commission's underlying policy of demonstrating and expanding its still-evolving federal powers vis-à-vis member states.

Simultaneous ratification, in fact, was a departure from customary practice even for the European Community. Individual states normally adhered to international treaties as soon as they completed their respective internal ratification processes, which might vary considerably in duration. The Commission's legal adviser, John Temple Lang, had noted in 1986 that "in practice it [the Community] usually accedes after most of its Member States have done so and not before."[23]

The attempt to impose an unaccustomed discipline of simultaneous ratification effectively tied the entire European Community's accession to the pace of its slowest member, which might be a small state with a low percentage of global CFC consumption. This policy precluded the large European CFC-consuming countries, whose ratification was essential for the treaty to become operational, from joining when they were ready. It also prevented a "bandwagon effect" and set a poor example for the rest of the world, which was waiting to see which major countries would ratify.[24] There was justifiable concern—echoed by the U.K. House of Lords—that the EC policy could postpone the protocol's entry into force.[25]

Not surprisingly, there was some confusion over this issue, and the EC Commission was prickly about criticism of its approach. In a letter to the editor of the *International Herald Tribune* nine months after the Montreal signing, EC environment director Laurens Brinkhorst reacted with irritation to an article by the chief U.S. negotiator referring to the possible delay involved in simultaneous ratification. Brinkhorst declared that "questioning of the united EC approach by a former [*sic*] U.S. official fundamentally distorts the environmentally beneficial effects of joint EC adhesion. Rather than slowing down it will accelerate the entry into force of the protocol."[26]

Illustrative of the muddle was a Commission document of December 1987 asserting that "Member States should not ratify either the [Vienna] Convention or the [Montreal] Protocol except simultaneously with the Community because if they were to do so . . . they would have to make a reservation concerning Community competence, and reservations are not permitted under either [treaty]."[27] Yet, at the time this statement was being written, two EC members, the United Kingdom and France, had already independently ratified the Vienna Convention.

Finally, weeks after Japan and the Soviet Union had ratified the protocol, and at the last possible moment to avoid a delay in the treaty's taking effect, the EC Commission ratified in mid-December 1988—together with 8 of its 12 member governments. Two states, Luxembourg and Portugal, had inadvertently ratified two months earlier; Belgium and France were unable to complete their domestic process until the last days of the year. The principle for which the Commission had delayed ratification by several countries had turned out to be substantively meaningless.

Thus, the Montreal Protocol did enter into force on January 1, 1989, with ratifications by 29 sovereign countries, plus the European Commission, together accounting for an estimated 83 percent of global consumption of CFCs and halons.

The Road to Helsinki

10

As calls multiplied for elimination of chlorofluorocarbons during the months following the Ozone Trends Panel report, the international chemical industry entered into a ferment of activity. In September 1988, the U.S. Alliance for Responsible CFC Policy announced support for a phaseout. Du Pont's early action was followed, after some delay, by voluntary decisions of other major U.S. and European companies gradually to cease producing CFCs. The United Kingdom's Imperial Chemical Industries, in its own policy reversal, asked for an urgent international review to strengthen the provisions of the Montreal Protocol aimed at phasing out CFCs (although there was no mention of halons).[1]

Challenges for Industry

During the second half of 1988, many other industries responded to the new science. Food packagers in the United States and the United Kingdom announced plans to stop using CFCs in the manufacture of disposable plastic-foam containers. The U.S. Polyurethane Foam Association, comprising producers of flexible foam, stated that they would introduce recycling and new processes to eliminate reliance on CFCs by the end of the century. In the United Kingdom, plastic-foam insulation manufacturers agreed to an early phaseout. American automakers voluntarily accepted new Environmental Protection Agency standards to permit increased use of recycled CFCs in automobile air conditioners, which now accounted for one-fourth of U.S. CFC consumption. Related developments were reported from the Federal Republic of Germany, Japan, and other major producing and consuming countries.[2]

By late 1988, it was generally accepted that a substantial total reduction

—at least 50 percent—in combined CFC and halon use could be accomplished relatively quickly and at little cost. Although much testing and product development still needed to be done, results from more than a year of intensive worldwide research indicated that there were several potential substitutes for CFCs that could become commercially viable.

Aerosols, which still accounted for about one-third of global CFC consumption, were obvious candidates for early virtual elimination (a small exception might be considered for certain unique pharmaceutical applications). Emissions from CFC 113 solvents in the electronics and other industries, which had grown to about 16 percent of worldwide consumption, could be cut substantially by a combination of substitutes and better containment and recycling practices. Japan, for example, had become particularly efficient in recovering over 95 percent of CFC solvents, in contrast to the United States, where there was much room for improvement. Similarly, large reductions in CFC use for plastic-foam production, which amounted to about one-fourth of global consumption, appeared technically feasible through recycling and substitution.

For refrigeration and air conditioning, however, representing 25 percent of the world's CFC consumption, feasible alternatives were not yet obvious—and this was the fastest-growing sector. There were also no chemicals with equivalent characteristics to halons for their specialized and important uses in fighting fires in aircraft, electronic equipment, oil rigs, nuclear power plants and vessels, and defense installations. However, confining halons to the most essential purposes, combined with eliminating such wasteful practices as spraying areas purely for testing, could bring some reductions.

Although substantial cutbacks in the controlled substances appeared feasible to most observers, there was controversy over prospects for their early elimination.[3] Some environmental groups minimized the costs and difficulties of a rapid phaseout.[4] Other analysts, however, noting that CFCs were employed in hundreds of manufacturing processes, raised concerns that replacements might be both less effective and more expensive. Researchers at Oak Ridge National Laboratory were "convinced that replacing these regulated CFCs represents . . . a major technological challenge."[5]

It was believed that some replacement technologies, particularly in refrigeration, would consume more energy—a result that would conflict with the goal of conserving energy to counteract greenhouse warming.[6] (An estimated 40 percent of U.S. usage of CFCs in 1985 was related to their energy efficiency characteristics.)[7] Another major problem was servicing existing equipment that had been designed to use CFCs. In the United States alone, an estimated 100 million refrigerators, 90 million air-conditioned cars and trucks, and tens of thousands of large air-conditioned commercial buildings already were using CFCs and could only gradually be replaced.

More Scientific Worries

Scientific developments in early 1989 added to anxiety over the fate of the ozone layer. In the winter of 1988–89, scores of scientists from the Federal Republic of Germany, Norway, the United Kingdom, and the United States mounted an Arctic expedition to investigate possible analogues to the Antarctic ice-crystal phenomenon. Although the scientists did not observe comparable ozone losses, they did find a "highly perturbed" Arctic atmosphere, with chlorine compounds present in concentrations 50 to 100 times greater than predicted. The mission concluded that the Arctic was poised for a potential ozone hole. The extent of future ozone depletion would depend in part on chance: specific weather conditions, particularly the timing of late-winter warming, would affect the potency of ice-crystal surface chemical reactions.[8] The implications of such a development for populous regions in northern latitudes were portended by reports from Australian scientists that the migration of ozone-poor air from Antarctica had caused seasonal declines in measured ozone concentrations of as much as 10 percent over populated parts of New Zealand and southern Australia.[9]

Researchers also began seriously to speculate whether similar stratospheric surface chemistry could intensify the ozone-destroying capability of chlorine even over areas where ice crystals caused by extreme polar climate conditions did not exist. It was suggested that a violent volcanic eruption could propel into the stratosphere minute sulfate particles, which laboratory studies indicated could provide surfaces for such chemical reactions. This hypothesis raised an apprehension that massive ozone losses—and consequent ultraviolet bombardment—could occur over heavily populated regions of the planet. A pioneer of the ozone-depletion theory, Ralph Cicerone, termed this new latent danger "a potential bombshell."[10]

The demonstrated inadequacy of existing models accurately to predict future ozone levels led scientists and, later, policymakers to place growing reliance on another concept: the total abundance of chlorine in the atmosphere, or "chlorine loading," as a measure of the potential threat to ozone.[11] It was also discovered that the calculated ozone-depletion-potential (ODP) value for a given chemical could be a misleading indicator of its impact on the ozone layer. Reduction strategies would henceforth increasingly be evaluated by their effect on chlorine loading rather than, as in 1987 and earlier, by the chancier model predictions of future ozone levels.

Total chlorine abundance in the atmosphere was estimated at three parts per billion (ppb) molecules of air in 1985—five times the level before CFCs had been emitted in large quantities. Accumulations of bromine, derived from the halons, also acted to deplete ozone: it was estimated that bromine was responsible for between 10 and 30 percent of the ozone loss over Antarctica, even though its global average atmospheric concentration amounted to only a fraction of one percent of chlorine, or about .02 ppb (20 parts per trillion).[12] However, chlorine became the key indicator because understanding of bromine loading and chemistry was still imperfect.

On the basis of the new concepts, analysts at the Environmental Protection Agency began to reexamine chlorine-containing substances that had not been serious candidates for regulation during the Montreal Protocol negotiations. The results were not reassuring. In terms of their contribution to anthropogenic chlorine concentrations as of 1985, the CFCs controlled under the protocol accounted for only about 64 percent. Two chemicals not regulated by the protocol, carbon tetrachloride (CT) and methyl chloroform (MC), each contributed 16–17 percent of the total chlorine loading attributable to industrial activity; indeed, each of them was responsible for four times as much chlorine as CFC 113.[13] (See Table 2.1 for a summary of characteristics of substances discussed in the context of the Montreal Protocol.)

Carbon tetrachloride was clearly in the same league as the CFCs: it has a relatively long atmospheric lifetime—50 years—and its ozone-depletion potential exceeds that of all CFCs. Its global consumption in 1986 was greater in tonnage than that of all the CFCs and halons combined. CT had been omitted from the earlier negotiations primarily because the full extent of its worldwide usage had been underestimated and it was considered almost impossible to control because it was cheap, simple to produce, and emitted from countless sources. Most Western nations had severely restricted CT because of its serious toxic and carcinogenic properties. In the United States, Western Europe, and Japan, CT was mainly employed as a chemical feedstock (notably for the production of CFCs), a safe process in which it was chemically transformed and caused no chlorine emissions; it was also used in small quantities in anticorrosive protective coatings for ships and bridges. However, carbon tetrachloride was also used in numerous small-scale operations throughout the developing world, Eastern Europe, and the Soviet Union as a commercial solvent, dry cleaner, pesticide, and grain fumigant.

Methyl chloroform had not been seriously considered for control under the Montreal Protocol because its short atmospheric lifetime (about six years) and its low ozone-depletion potential (only about one-seventh that of CFCs 11 and 12) had made it seem relatively innocuous at the time. MC was, however, widely used in industrial economies as a solvent for precision and metal cleaning as well as in adhesives and coatings. In fact, later research revealed that its 1986 consumption, in tonnage, exceeded that of both CFC 11 and CFC 12. Use of MC was pervasive throughout the manufacturing sector, particularly in the electronics, aerospace, and automotive industries. Most businessmen were unaware of the threat it posed to the ozone layer, and many were planning to substitute it for the regulated CFCs.

In 1987 it had seemed neither necessary nor practical to delay an agreement on the more dangerous CFCs and halons for the sake of trying to negotiate methyl chloroform and carbon tetrachloride into the treaty. Now, however, serious consideration would have to be given to adding them to the protocol's controlled list.

In addition, EPA began to look more critically at another family of chemicals related to the CFCs—the hydrogen-containing halocarbons. These hydrochlorofluorocarbons (HCFCs) and hydrofluorocarbons (HFCs) were now under development by the chemical industry as promising alternatives to CFCs. As the HFCs contain no chlorine and hence pose no threat to the ozone layer, they would not be subject in any case to regulation under the Montreal Protocol; however, it was discovered that their global warming potential was significant. With respect to the HCFCs, the new studies suggested that if their use as substitutes were greatly expanded, they could delay a decline in stratospheric chlorine concentrations even as CFCs and halons were being phased out.[14] These chemicals, then, could no longer be regarded as panaceas, but only as interim solutions pending another yet-undiscovered generation of substitutes and technologies.

London: The South Speaks Out

These developments added urgency to the widening consensus that stronger international action was needed to protect the ozone layer. On the eve of the conference on ozone convened by U.K. Prime Minister Margaret Thatcher in early March 1989, both the European Community and the U.S. government announced formal commitment to a full phase-out of CFCs by the year 2000. An important difference still prevailed: the United States specifically advocated a halon phaseout, while the EC remained silent on this subject.

Evidencing the growing international public involvement, which had been absent during the actual protocol negotiations except in the United States and a handful of other countries, representatives of some 93 environmental organizations from 27 nations, including developing countries, met in London just before the governmental conference. They issued a joint statement urging an aerosol ban by 1990 and a total phaseout of CFCs and halons, plus carbon tetrachloride and methyl chloroform, no later than 1995.[15]

The official London Conference on Saving the Ozone Layer, cosponsored by the U.K. government and UNEP, proved an important political milestone. It effectively set the stage for the First Meeting of Parties to the protocol, which would take place two months later in Helsinki. Delegates from 123 countries, of which 80 were represented at the ministerial level or equivalent, produced an impressive political expression of global concern over the ozone layer and overwhelming support for the Montreal Protocol. There was a consensus that the "ultimate objective" had to be total elimination of CFCs and halons; this was the same phrase that only 18 months earlier had been unacceptable to the United Kingdom and the EC Commission except in a preambular clause to the protocol. Many countries also announced in London their firm intention to accede to the treaty.[16] As hoped, a major result of the London conference was to increase the pace of government ratifications: by the end of 1989, an additional 18 nations—of which 14 were developing countries—became parties to the treaty.

In contrast to the situation during negotiation of the protocol, wide agreement now prevailed on the science as well as on the desirable control measures. However, the treatment of developing countries, which had ostensibly been resolved at Montreal, reemerged in London as a central issue.

Governments of developing nations at London noted that products made with or containing CFCs, especially for food preservation and air conditioning, were essential to raising their living standards. Chinese and Indian delegates stated that it was unacceptable for developing countries either to have to forgo these necessities or to pay more for substitutes and thereby further enrich the very chemical industries that had created the ozone problem in the first place. The Indian environment minister declared that what haunted developing countries was the prospect that substitutes would not become available soon enough to meet rising expectations or that the alternatives would prove too expensive.[17]

Thus, the developing nations at London demanded assurances that financial aid and technology transfer would be forthcoming. For their part, industrialized countries now realized that the promises of Montreal would have to be translated into tangible assistance to enable developing countries to forgo significantly expanded use of CFCs.

Helsinki: The Treaty Is Launched

Against this background, the First Meeting of Parties to the Montreal Protocol was convened by President Mauno Koivisto of Finland in Helsinki on May 2, 1989. The treaty, which had entered into force only five months previously, could not legally be revised at this first meeting. The Vienna Convention (article 9, paragraph 2) and the Montreal Protocol (article 2, paragraph 9) required that any proposed amendments and adjustments of the protocol be submitted to UNEP for communication to all parties at least six months before the meeting of parties that would consider them. Nevertheless, the Helsinki conference produced a political agreement by the parties to strengthen international controls plus a detailed process for accomplishing this within the coming year.

The contrast between the Helsinki meeting and the negotiations leading up to Montreal could not have been more striking, in terms of breadth and level of participation, sense of urgency, and extent of consensus. It was difficult to believe that a scant two years earlier it had been difficult to assemble more than 25 to 30 delegations, mostly from the industrialized world, to discuss the arcane subject of stratospheric ozone.

At Helsinki, delegates from over 80 nations were in attendance, most of them from developing countries. About one-third of the governments were represented at ministerial or deputy-ministerial rank, including Belgium, China, the Federal Republic of Germany, the German Democratic Republic, Hungary, Japan, Malaysia, Mexico, New Zealand, Norway, the Soviet Union, Spain, Sweden, the United Kingdom, and Zambia. Many countries were present for the first time at a negotiation dealing with a global environmental threat. There were also observer delegations from a wide range of industrial associations and environmental organizations, plus more than 200 correspondents from the international media.

Mostafa Tolba again set the tone in his opening address. He stressed the evidence that stratospheric ozone was being depleted more rapidly than had been anticipated when the treaty was negotiated, and noted the clearer understanding that CFCs could also significantly add to greenhouse warming. Tolba called for elimination of all ozone-depleting substances by the end of the century and establishment of an international fund to generate, subsidize, and transfer new technologies and products to the developing countries.[18]

On the first point, there was extraordinary unanimity. EPA made a special presentation to the conference of its latest estimates of future chlorine loading of the atmosphere under various assumptions of emissions of CFCs, halons, carbon tetrachloride, methyl chloroform, and HCFCs. Governments that two and a half years earlier had resisted going beyond a freeze of only CFCs 11 and 12 now joined in calling for early phaseout of a broad range of chemicals. Numerous African, Asian, Eastern European, and Latin American countries spoke for the first time in strong support of such measures. Several delegations specifically recommended adding carbon tetrachloride and methyl chloroform to the list of controlled substances.

At the initiative of the government of Finland, the meeting of parties approved by consensus the Helsinki Declaration on the Protection of the Ozone Layer. This was a nonbinding document, but it carried important political weight for the coming deliberations on strengthening the protocol. It called for phaseout of CFCs "as soon as possible but not later than the year 2000." Reflecting somewhat less agreement on timing, but nevertheless a substantial change since Montreal, the governments also pledged in the declaration both to "phase out halons and control and reduce other ozone-depleting substances which contribute significantly to ozone depletion as soon as feasible." (The latter was a reference to carbon tetrachloride and methyl chloroform, as well as HCFCs.) Details of these reductions would obviously require negotiation in the months ahead.

Concerning Tolba's second major recommendation, there was wide recognition at Helsinki that financial and technical assistance for developing countries was essential. The creation of an international fund for this purpose was publicly supported by many developing nations, as well as by Finland, the Netherlands, New Zealand, and Norway. Norway expressed willingness to commit 0.1 percent of its gross national product to an international fund for protection of the atmosphere (including climate change as well as the ozone layer), but major donor countries did not respond to this idea. The EC, Japan, and the United States were unwilling to commit themselves to a new institution and insisted that the actual needs and the possibilities for using existing bilateral and multilateral assistance channels had to be carefully examined.

A compromise decision established a special working group to develop recommendations for the next meeting of parties on "adequate international funding mechanisms which do not exclude the possibility of an international Fund."[19] The debate on this issue, though energetic, was characterized by a lack of political rhetoric and by a clear understanding that effective ways had to be found to ensure that developing countries would not be disadvantaged by the phaseout of CFCs.

In addition, the parties at Helsinki considered another important developing-country issue: a clarification of the article 5 provision for low-consuming countries to expand their use of the controlled substances, up to a specified ceiling, in order to meet their "basic domestic needs." After the protocol was signed, some developing-country governments had argued that "basic domestic needs" included the "need" to increase exports. The Helsinki participants, however, decided that the expanded manufacture of products containing CFCs solely for purposes of supplying other nations (as opposed to satisfying domestic demand) was inconsistent with the intent of the protocol to limit global use. Consequently, an agreement was reached to exclude increased exports from the permissible growth in developing countries' use of CFCs.

An unexpectedly intense discussion developed in Helsinki over data confidentiality. This debate evoked memories of the exchanges at Montreal over EC claims for special privileges as a regional economic integration organization (REIO). At the October 1988 meetings in The Hague, participants had been unable to agree on how production, import, and export data should be reported to UNEP. The EC had contended that its member countries need not report separate national data and that the Community would instead submit aggregated data for all 12 nations. Further, the EC had argued that the chemicals need not be reported individually but rather as a combined "basket," since the controls applied to a basket of CFCs rather than to each separate chemical. The EC had maintained that because some EC members contained only one or two companies (in contrast to the United States and Japan), release of disaggregated production data on a country-by-country basis would reveal too much to competitors.[20]

Other countries at The Hague, including New Zealand and the United States, had countered that, as long as REIO member governments maintained separate status as parties, they had to submit separate national data. Moreover, notwithstanding that the controls applied to "baskets," the protocol was very specific in requiring reporting of individual chemicals (article 7). The EC was reminded by other delegations that this degree of specificity was necessary not only to aid scientists in their modeling exercises but also to monitor compliance with the treaty.

It was difficult to believe that, with so few major companies in the business, each one did not have a good estimate of the others' output, quite apart from UNEP reporting. The real issue seemed to be access to the data by other governments and by concerned citizens' groups.

At Helsinki, the EC and the United States jointly proposed a compromise, stipulating that detailed data would be reported to UNEP but that UNEP would reveal such data to the public only in aggregated form in those cases in which a given reporting party desired such confidentiality. All data, however, would be available on request to other governments for examination on a confidential basis. Dissemination of disaggregated data to the general public was thereby precluded.

Although the compromise proposal was eventually accepted at Helsinki, it was only after objections had been expressed by an unusual coalition of speakers, including New Zealand, Nigeria, Norway, the Soviet Union (which thereby reversed its pre-Montreal position), Zambia, Greenpeace, and World Wide Fund for Nature. Speakers noted that the EC's continuing insistence on confidentiality made it difficult to avoid the impression that some EC member states were still trying to preserve advantages for their chemical companies. Many participants felt that the preoccupation with preserving secrecy for a declining business sector was inappropriate in light of the desirability of public knowledge as a deterrent to possible attempts to evade required production cuts.

In addition to considering these issues, the First Meeting of Parties clarified several ambiguous protocol provisions and set up a special legal group to develop procedures for determining and dealing with noncompliance. It also established work plans for research and development, exchange of information, technical assistance, and public education efforts.

A major accomplishment at Helsinki was the elaboration of a detailed timetable and mechanism for implementing the protocol and strengthening it at the Second Meeting of Parties, scheduled for June 1990 in London. Central to this process was the creation of an Open-Ended Working Group of the Parties to the Montreal Protocol, which, along with spinoff legal and technical working groups, would meet many times over the ensuing months to prepare for the crucial 1990 conference. The working groups would be open to participation by any government, including those not yet parties to the treaty, as well as nongovernmental observers; to encourage attendance by representatives of developing countries, UNEP and some governments offered financial aid.

Key to the deliberations of these working groups would be the findings of the scientific, environmental, technological, and economic expert panels that had been established in October 1988 at The Hague meeting, even before the protocol had officially entered into force. The work of these panels was well advanced and would be completed in the next few weeks.

A science panel was reviewing and analyzing the implications of the most recent measurements and predictive models of stratospheric ozone trends, as well as updating calculations of ozone depletion, chlorine loading, and global warming potentials for various chemicals—including those being developed as substitutes for CFCs. The environmental panel was assessing the state of knowledge of health and environmental effects of altered ozone concentrations and increased ultraviolet radiation. A technology panel, which included many industry representatives, was analyzing technical options to determine and quantify the feasibility of reducing ozone-depleting substances. And, finally, the economics panel was examining the benefits of reduced use of CFCs and halons, the costs of technical solutions, and the implications for technology transfer to developing countries.

In conclusion, participants at Helsinki were encouraged that the deliberations generally drew on the protocol's spirit and intent for guidance rather than seeking legalistic loopholes to dilute the force of the treaty. The Montreal Protocol was thereby well and truly launched.[21]

The Protocol in Evolution

11

The four assessment panels completed their work in the summer of 1989. This undertaking, which included peer review, involved nearly 500 scientists and other experts from around the world and produced over 1,800 pages of findings. The integrated summary of the panels' conclusions—known as the *Synthesis Report*—became the basis for the negotiations of the Open-Ended Working Group that had been created in Helsinki to prepare recommendations for revising the Montreal Protocol.

The *Synthesis Report:* Restoring the Ozone Layer

The discussions of the Open-Ended Working Group would prove to be dominated by the findings of the science panel. The scientists had become increasingly concerned about the inability of their models to predict accurately the actually measured depletion of the ozone layer. Their misgivings were heightened by the new elements in the science since Montreal: the unexpectedly large ozone losses measured over northern latitudes, the effects of the annual Antarctic ozone hole on southern populated latitudes, the potential for precipitous ozone depletion over the Arctic, and the ominous new factor of possible consequences from volcanic eruptions.

The chlorine loading methodology was critical to the science reassessment. The preindustrial level of atmospheric chlorine was calculated at 0.6 part per billion (ppb), representing methyl chloride, a naturally occurring chemical. The Antarctic ozone hole had begun to develop during the 1970s, when global average chlorine concentrations had climbed to the 1.5–2 ppb range. Chlorine loading had reached 3 ppb by 1985 and was expected to rise to 3.5–3.6 ppb by 1990.[1]

A logical benchmark for evaluating future control strategies was the return of atmospheric chlorine concentrations to no higher than 2 parts per billion—roughly the chlorine loading at which Antarctic springtime ozone levels had begun to drop sharply in the late 1970s. According to new calculations, chlorine abundance under the existing control provisions of the Montreal Protocol could reach an almost inconceivable 11 ppb, nearly 20 times natural levels, by the end of the next century—and still be in an upward trajectory (see Figure 11.1). Even tightening the measures in order to stabilize present chlorine concentrations, which dur-

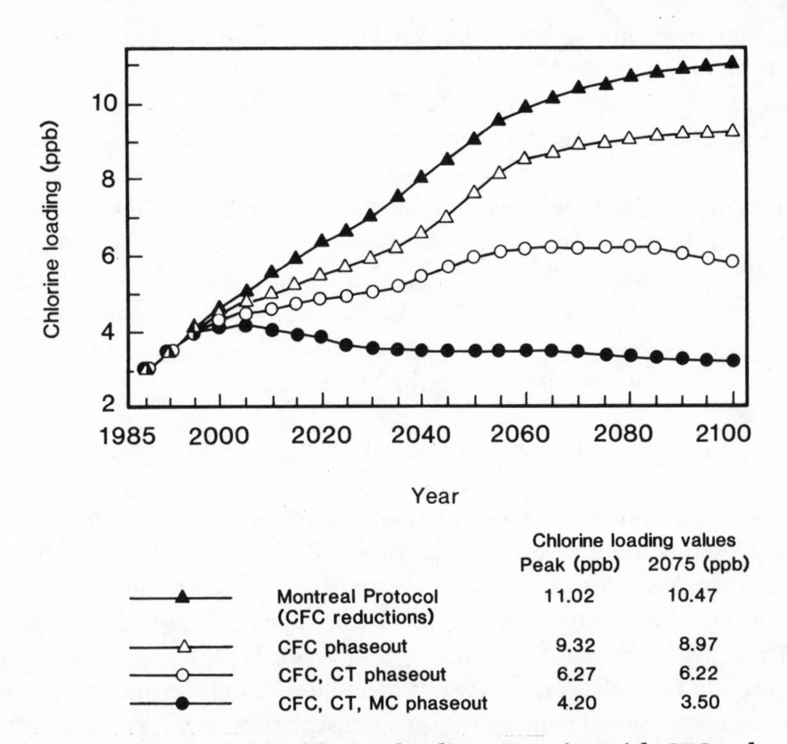

Figure 11.1 Atmospheric chlorine-loading scenarios with CFC reductions and CFC, CT, and MC phaseouts, 1985–2100. (Assumptions for all scenarios: intermediate reductions as specified in Montreal Protocol or London revisions; HCFC substitutes replace 30 percent of CFC reductions and have an atmospheric lifetime of 8 years; no controls on future HCFC use; natural background chlorine loading of 0.6 ppb.) *Source:* Calculated by EPA.

ing the Montreal negotiations had seemed an audacious proposition, would condemn Antarctica to a recurring ozone hole in perpetuity—with unclear implications for the rest of the planet.

The science review indicated that recuperation of the atmosphere from its human-imposed chlorine binge would take decades and would require drastic near-term action. It was now realized that even if all chlorofluorocarbons were phased out, chlorine concentrations would continue to increase steadily as a result of chlorine loading from carbon tetrachloride and methyl chloroform. As illustrated in Figure 11.1, if CT and MC were also phased out, peak chlorine would be lower, but the 2 ppb benchmark would still not be reached because expanding uses of hydrochlorofluorocarbon substitutes—even if for only 30 percent of the banned chemicals' markets—would keep the chlorine level up.

In fact the science assessment demonstrated that ozone layer recovery would require early elimination of all CFCs, plus CT and MC, together with only transitional reliance on HCFC substitutes. And even in this case, as a result of the existing stock of long-lived CFCs in the atmosphere, the 2 ppb target could not be reached for at least 80 years—until around 2070 to 2080, depending on assumptions. Moreover, the chlorine level would continue to rise before it began a gradual, decade-long descent. (Figure 11.2 illustrates this phenomenon under three different assumptions regarding the characteristics and duration of use of HCFC substitutes.) Chlorine loading was expected to increase from its existing level to something over 4 ppb even under the most optimistic assumption of future controls. Unquestionably, there would be additional depletion of the ozone layer before it could begin to recover.

The environmental effects panel updated what had been published in the comprehensive 1986 EPA study (see Chapter 2). With the exception of future rates of skin cancer and eye cataracts, it remained impossible to quantify effects with any degree of confidence—including harm to the human immune system and to agriculture, fisheries, forests, natural ecosystems, air quality, and materials. The threats to animal, fish, and plant life from substantially increased ultraviolet radiation during the Antarctic springtime constituted a new area of uncertainty.[2]

There was still no evidence of increased biologically damaging ultraviolet radiation (UV-B) that would induce the manifold harmful effects. On the contrary, measurements continued to indicate a decrease in UV-B reaching Earth's surface.[3] However, while the total quantity of protective ozone throughout the atmosphere as a whole had indeed declined, a redistribution had taken place, with greater ozone concentrations at low altitudes somewhat offsetting ozone losses in the stratosphere. Scientists postulated that low-altitude ozone—created mainly by pollution from transport and industry—was more efficient in absorbing UV-B and therefore made up for the decline in stratospheric ozone.[4] One form of pollution was, in effect, canceling out another; but as stratospheric ozone continued to fall, and as planned measures to improve air quality began to reduce noxious low-level ozone, the predicted increase in UV-B at Earth's surface would become evident.

On balance, scientific experts were clearly worried about the uncertain consequences of postponing restoration of the atmosphere to its earlier

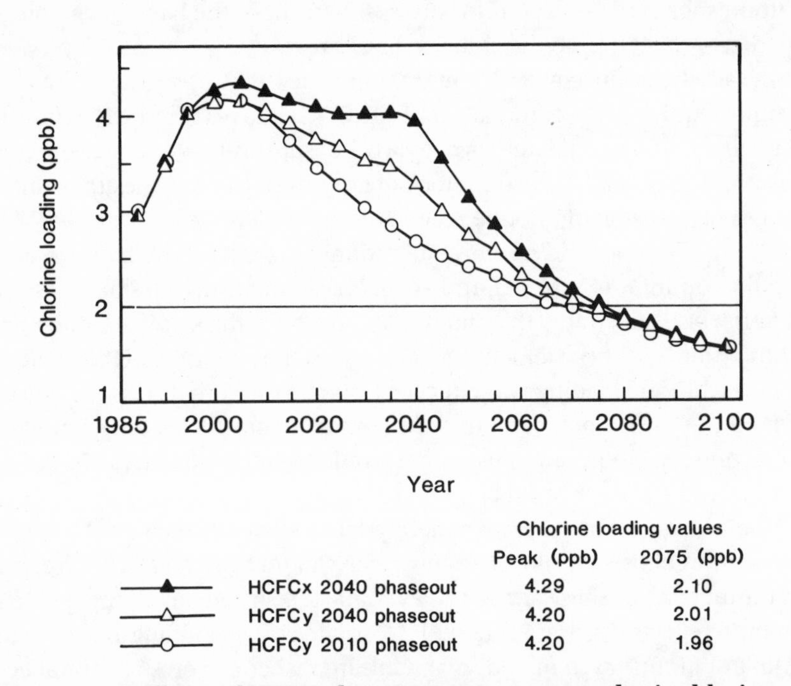

Figure 11.2 Effects of HCFC phaseout year on atmospheric chlorine loading, 1985–2100. (Assumptions for all scenarios: phaseout of CFCs, CT, and MC as specified in London revisions; HCFC substitutes replace 30 percent of CFC reductions; HCFCx substitutes have an atmospheric lifetime of 15 years, chlorine-loading potential of 0.13; HCFCy substitutes have an atmospheric lifetime of 8 years, chlorine-loading potential of 0.05; natural background chlorine loading of 0.6 ppb.) *Source:* Calculated by EPA.

condition. The *Synthesis Report* declared that a complete and timely phase-out of all major ozone-depleting substances, with worldwide compliance, was "of paramount importance in protecting the ozone layer."[5] And the panel stressed that the longer the delay in implementing such measures, the longer would be the recovery time for the ozone layer. Because of the nonlinear response of polar ozone to increased chlorine concentrations over Antarctica, the possibility could not be ruled out that a similar unforeseen threshold could be crossed over the Arctic region if chlorine levels continued to rise. Mostafa Tolba expressed this concern in his report to the first meeting of the Open-Ended Working Group in Nairobi in August 1989: "The uncertainty regarding how the atmosphere responds to human intervention is itself sufficient reason for resisting its alteration and if possible returning it to its pre-industrial state."[6]

The accelerating pace of industrial research, testing, and technological innovation in 1988 and 1989 had considerably increased the options for replacing and conserving CFCs. The technological assessment panel concluded in its report to the working group that it was now "technically feasible" by the year 2000 to phase down the five controlled CFCs by at least 95 percent, methyl chloroform by 90–95 percent, and carbon tetrachloride by 100 percent.[7] However, the panel could not agree on a feasible phaseout date for halons because of the continuing unavailability of prospective substitutes. A majority of experts felt that conservation and wider application of existing fire-protection techniques could permit an orderly elimination of halons by 2005, but others believed that any reduction schedule would be premature until more information was available on essential needs and possible alternatives.

The economic assessment panel identified, but could not quantify, "enormous beneficial impacts on human health and the environment" from cutting back on ozone-depleting substances. It concluded that "the monetary value of the benefits of safeguarding the ozone layer is undoubtedly much greater than the costs of CFC and halon reductions."[8] With respect to costs involved in accelerating the protocol's reduction schedule, much would depend on the rate of technological progress, which was impossible to predict. Such costs would include research and development, testing substitutes for safety and toxicity, capital investment, and possibly higher operational outlays. The economics panel concluded that very rapid phaseout ("much less than 10 years") would entail substantially higher costs primarily as a result of premature abandonment of capital goods and sectoral unemployment.[9] Past experience indicated, however, that continued rapid innovation might well result in lower costs than those projections based on existing technologies.

Industry Faces the Inevitable

Reactions to the growing alarm of the scientific community varied somewhat according to perspective. Most of industry was by this time generally reconciled to the inevitability of eventual elimination of all substances that threaten the ozone layer. However, the large chemical companies and some particular industry branches, such as the solvents sector, stressed the costs and difficulties of phaseout and appeared to want to postpone it as long as possible.[10]

The chemical manufacturers maintained that the big phaseout cost factor would be neither the development of alternatives nor the price of substitute chemicals, even though these were estimated to be several times more expensive than CFCs. Rather, the largest expenditures would fall upon user industries, which would be obliged to redesign existing equipment and develop new processes to accommodate the different characteristics of the CFC replacements; the major element would not be the cost of new types of refrigerants, for example, but the costs of designing and producing a completely new refrigeration unit and perhaps prematurely scrapping both the old unit and the equipment used to make it. Du Pont estimated that, worldwide, as much as $385 billion in capital equipment was dependent on CFCs; for the United States alone the figure was $135 billion.[11]

Because of the many variables, there were no firm estimates from industry of the magnitude of these costs, although some sources spoke in terms of scores of billions of dollars. An AT&T spokesman may have been close to the mark in noting that although "80 percent of the solution comes fairly easy . . . the last 20 percent gets very expensive and difficult to achieve."[12] A representative of the Society of Automotive Engineers suggested that in the automobile sector alone, the cost of retrofitting 140 million American cars by 1997 to use non-CFC-cooled air conditioners could range from $70 billion to $140 billion. Environmentalists derided this estimate for ignoring the possibilities for "drop-in" CFC replacements, recycling CFCs to preserve existing air conditioners until their normal obsolescence, or both. For its part, Imperial Chemical Industries (ICI) was emphatic that it would be "virtually impossible" to come up with drop-in substitutes for coolants.[13]

Other sources were less negative. EPA estimated the costs of a CFC phaseout in the United States at about $3 billion, mostly for retrofitting or replacement of capital equipment.[14] There was also concern in some quarters that research was overly dominated by the existing CFC manufacturers, which had a vested interest in developing patentable related chemical alternatives. Some researchers maintained, for example, that in the critical refrigeration sector there were feasible—but less profitable—alternatives to the particular chemicals and processes being promoted by the big chemical firms.[15] Even in user industries there appeared some unease about overreliance on the chemical producers as CFCs were phased out, and many firms embarked on their own ambitious research programs.[16] AT&T, which in 1988 had pioneered the development of a citrus-based alternative to CFC 113, stated that it would eliminate CFCs by 1994; its new circuit board plant in Singapore was the first in the industry to use no CFCs. Northern Telecom announced that it would end reliance on CFCs by 1991, Motorola set a date of 1992, and International Business Machines fixed on 1993. Boeing targeted a 60 percent reduction by 1992, 90 percent by 1999. General Motors and Nissan developed machinery to recapture and reuse CFCs from automobile air conditioners. Carrier Corporation came up with a new recycling system for commercial refrigeration units. Polyurethane-foam makers reported that they could meet any revised CFC phaseout schedule.[17]

The chemical industry estimated in 1989 that about 30 percent of existing and growing demand for CFCs could be replaced relatively inexpensively merely through improved conservation, recovery, and recycling.[18] This was a rather startling admission, considering that only two years earlier much of the industry—particularly in Europe—had insisted that any CFC reductions would have severe economic consequences. An additional 30 percent of demand could be met through substances or processes that did not release chlorine emissions; for example, by replacing CFCs in aerosol sprays, cleaning agents, and plastic-foam blowing.

However, for the remaining 40 percent of demand, in refrigeration, air conditioning, certain solvents, and manufacture of insulating foams, the chemical producers insisted that the only feasible alternatives in sight were the hydrochlorofluorocarbons (HCFCs), accounting for 30 percent, and hydrofluorocarbons (HFCs), accounting for 10 percent. Since the HFCs, which do not affect the ozone layer, were outside the scope of the Montreal Protocol, attention focused on the HCFCs.

Risks during Transition

Hydrochlorofluorocarbons break down in the atmosphere much faster than their CFC cousins. Hence, at such time as HCFC emissions were halted, their contribution to chlorine loading would fall rapidly and the ozone layer would benefit correspondingly quickly.

However, different HCFCs under study as possible substitutes ranged in their chlorine-loading and ozone-depletion potentials from 2 percent to as much as 14 percent those of the principal CFCs, while their atmospheric lifetimes varied from 2 to 20 years (see Table 2.1). Obviously, from the scientific standpoint, it would be desirable to choose among the HCFCs and concentrate on those with a substantially smaller impact on the ozone layer. But even if one assumed substitution by HCFCs at the lower end of these scales, their use over an extended period could prolong until the twenty-second century, or perhaps even indefinitely, the point at which chlorine concentrations would return to pre–ozone hole levels (see Figure 11.1).[19]

The *Synthesis Report* observed, however, that if the HCFCs were themselves phased out sometime between 2030 and 2045, they would have "little influence" on the date by which chlorine loading could fall to 2 ppb.[20] That target date of 2070–2080 was linked primarily to the deadline of 2000 for phasing out the CFCs. However, the longer a phaseout of HCFCs—especially those with high chlorine-loading potentials—was delayed, the longer it would take to reduce chlorine levels.

European chemical producers were adamant against including HCFCs in the Montreal Protocol—even for purposes of reporting data on production, exports, and imports. They based their argument on the substantially lower damage to the ozone layer from HCFCs in comparison with CFCs, while ignoring the more subtle long-term analysis of the science panel and the *Synthesis Report.* Britain's ICI, for example, paid lip service to an "eventual" phaseout of all ozone-depleting substances, noting that "research and development in the longer term may eventually provide us with a better solution than HCFCs." But ICI seemed to prefer some unspecified date in the hazy future, asserting that "if HCFCs are designated as Controlled Substances in the revised Protocol . . . the chemical industry will have no incentive to invest in production of these products."[21]

American industry had a different view. The Alliance for Responsible CFC Policy, which included both producer and consumer industries, acknowledged that "HCFCs are transitional substances and establishment of prudent HCFC phaseout dates is therefore encouraged." The Alliance maintained that the assurance of "normal product lifetimes" would provide greater incentives for industry both to invest in the HCFCs as "bridging chemicals" and to look further ahead to development of chlorine-free substitutes. Unlike the European chemical manufacturers, U.S. industry favored "time certain" phaseout dates in the revised protocol, arguing that uncertainty about regulation could cause both producer and user industries to delay investment commitments in the new products. Specifically, the American industry association advocated establishing phaseout dates within the period 2030–2050, with longer-lived HCFCs phased out in the earlier part of the range.[22] This position was fully consistent with the findings of the science panel.

The main concern of American industry was that the phaseout schedule should reflect the 30-to-40-year lifetimes of the equipment using HCFCs, such as industrial and commercial cooling units. The CFC Alliance stated that hundreds of millions of dollars had already been invested in research and development of HCFCs, but that the additional several billion dollars required over the next decade to bring the substitutes to market would not be forthcoming if phaseout dates were premature.

The Alliance contended that overly strict deadlines for eliminating HCFCs would not only slow their development by producers but also inhibit their early substitution for CFCs by users. There would be a temptation—particularly in developing countries eligible for the 10-year moratorium—to rely longer on CFCs and to wait for the ultimate nonchlorine technology rather than invest in equipment for too-transitory substitutes and be forced to make a costly second transition before full value of the HCFCs had been obtained. The U.S. industry group calculated that as little as one percent noncompliance with a CFC phaseout by high-growth developing-country economies would contribute more total chlorine to the atmosphere than 10 years of HCFC use by the entire world.

With respect to methyl chloroform, industry appeared prepared to make a stand despite the science panel's findings. U.S. companies produced over half of the world's total of this important solvent for the metals and electronics industries, with Dow Chemical alone responsible for 40 percent; over 73,000 American firms used MC.

A U.S. industry association, the Halogenated Solvents Industry Alliance, warned of the high costs of an MC phaseout, entailing "business shutdowns, loss of competitiveness and job layoffs."[23] Both Dow and ICI distributed letters to customers in 1989 urging an antiregulatory lobbying campaign with the U.S. and U.K. governments. Industry contended that MC's relatively low ozone-depleting potential and short atmospheric lifetime, its usefulness in "strategically important industries," its declining sales trend, and improved technologies for recovery and recycling all combined to make strong controls unnecessary.[24]

However, MC contributed about 16 to 17 percent of current anthropogenic chlorine loading of the atmosphere—four times as much as the soon-to-be-banned CFC 113. The science panel calculated that MC alone accounted for 0.5 ppb of total atmospheric chlorine concentrations.[25] Therefore, precisely because of its short atmospheric lifetime, an early phaseout of MC would bring the most immediate reductions in chlorine loading and benefit to the ozone layer. If MC emissions were merely frozen rather than phased out, the chlorine level would remain 0.5 ppb higher, and its rate of decline would be substantially slowed (see Figure 11.3).

Environmentalists were concerned that if MC were not subject to strict regulation, it could find growing markets as a replacement for CFCs in aerosols and electronics cleaning applications, which would then tie these users to an ozone-depleting chemical even as less harmful alternatives were becoming available.[26] The Natural Resources Defense Council, a U.S. environmental organization, estimated that MC was a component of hundreds of consumer products, mostly in aerosol sprays and many of them misleadingly labeled as "ozone friendly" on the disingenuous grounds that they did not contain the already-controlled CFCs.[27] New data substantially undermined industry's argument against stringent controls: it appeared that, far from declining, worldwide sales of methyl chloroform had actually grown by 30 percent from 1982 through 1988—rising by 8 percent in 1988 alone.[28]

These were among the new considerations that influenced government negotiators in the last half of 1989 and the first half of 1990 as they prepared to revise the Montreal Protocol.

Back to the Bargaining Table

In the months following the Helsinki meeting of parties, no less than seven formal meetings of the Open-Ended Working Group took place: in Nairobi in August 1989, in Geneva in September and November 1989 and in February, March, and May 1990, and in London immediately before the Second Meeting of Parties in June 1990. These sessions were supplemented by over a dozen smaller meetings and informal intergovernmental consultations, involving diplomatic, environmental, financial, legal, and technical experts.[29] A small mountain of documentation was produced as the governments focused on two fundamental and related is-

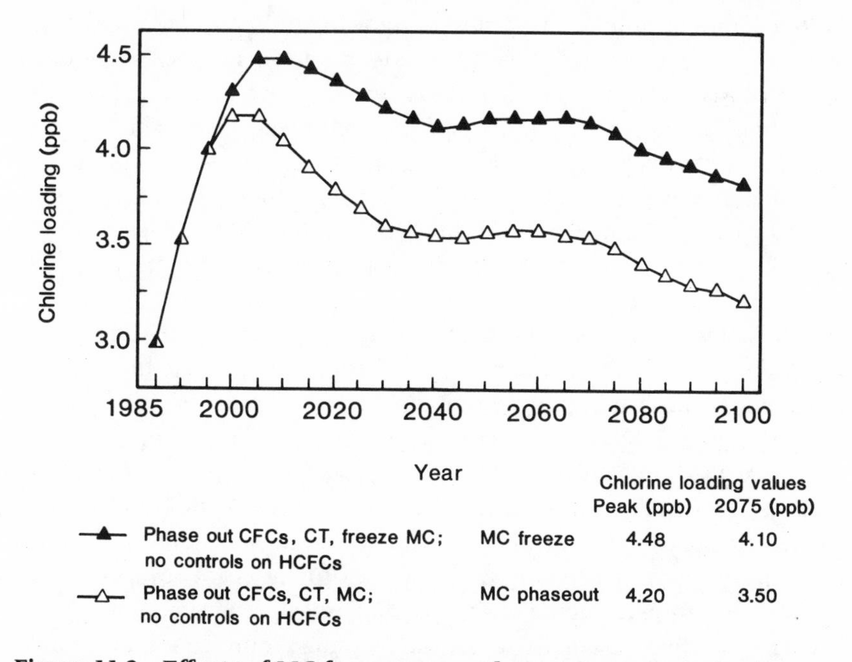

Figure 11.3 Effects of MC freeze versus phaseout on atmospheric chlorine loading, 1985–2100. (Assumptions for both scenarios: intermediate reductions as specified in London revisions; HCFC substitutes replace 30 percent of CFC reductions and have an atmospheric lifetime of 8 years; no controls on future HCFC use; natural background chlorine loading of 0.6 ppb.) *Source:* Calculated by EPA.

sues: strengthening the controls over ozone-depleting chemicals, and treatment of the developing countries.

After approving the seminal *Synthesis Report,* the Open-Ended Working Group set to work attempting to design changes to the treaty for consideration by the Second Meeting of Parties. The group generally devoted alternate meetings to the control provisions and the developing-country issues. In contrast to the negotiations leading up to Montreal, the working group sessions through May were usually attended by up to 50 government delegations, including such influential nonparties as Brazil, China, India, and South Korea; more than 80 governments participated in the crucial final working group meeting in June.

No distinction was made in the debates between delegates from parties and those from potential parties to the protocol. Only parties, however, were eligible to propose new language for the protocol; on a few occasions, representatives of noncontracting governments were reminded that, unless a party could be found to sponsor their proposal, it could not be formally included in the draft protocol texts being developed.

Industry was particularly well represented at the sessions working on the control measures, with observer delegations from at least a dozen industry groups ranging from the U.S. Motor Vehicles Manufacturers' Association to the Japanese Association for Hygiene for Chlorinated Solvents; Friends of the Earth International and Greenpeace International were usually the only environmental organizations in attendance.

Although these negotiations had approximately the same nine-month duration as those preceding the original treaty, there were many more meetings, and they were in many ways more complicated and difficult, reflecting the much higher number of participants and the increasing complexity of the issues. The broad consensus of the Helsinki Declaration proved unexpectedly difficult to render into legally binding provisions. Indeed, so many widely varying proposals for protocol revisions were presented that the evolving draft texts threatened to become unmanageable. It soon became apparent that the working group would not be able to agree on specific recommendations that, according to the terms of the Vienna Convention and the Montreal Protocol, had to be delivered by the UNEP secretariat to governments six months before a meeting of parties.[30]

With unaccustomed indulgence, government legal experts determined that this requirement did not mean that negotiators could not look at proposals presented after December 1989, "as long as such proposals are sufficiently related to proposals communicated to the Parties [under the six-month rule] . . . so that the Parties may be deemed to have had adequate opportunity to consider [their] merits."[31] This tolerant interpretation proved indispensable to the treaty revision process.

Accordingly, the working group's recommendations were submitted in December in the form of a single text with dozens of bracketed options and alternative formulations; governments were thus able to continue the necessary deliberations and even alter the proposed revisions during the period right up to the June 1990 meeting of parties.

Strengthening the Controls

In introducing the subject of stronger control measures to the working group in August, Mostafa Tolba offered uncharacteristically cautious opening "recommendations," not departing far from the vague Helsinki Declaration goals. Tolba's proposals included (1) a phaseout of currently controlled CFCs by the end of the century; (2) a 50 percent reduction of currently controlled halons by 1995, with an unspecified target date for their phaseout; (3) the inclusion of carbon tetrachloride and methyl chloroform as controlled substances, with unspecified phaseout schedules; and (4) a requirement to report data on production and trade in HCFC substitutes, plus an injunction that their use be subject to "careful limits."[32]

A month later, the Bureau of the Parties to the Montreal Protocol, a steering group elected by the parties and chaired by Finnish Environment Minister Kaj Bärlund, with high-ranking representatives from Kenya, Mexico, New Zealand, and the Soviet Union, met with Tolba in Geneva and formulated more-detailed recommendations for consideration by the working group.[33] Use of the bureau in this manner also had no analogue in the less complex process leading up to Montreal. The bureau convened with Tolba between formal working group meetings four times in 1989 and 1990 and played an important role in defining issues for the larger group.

The bureau's proposals to the Open-Ended Working Group on new control measures were as follows:

CFCs: 50 percent reduction by 1994 or 1995, 85 percent reduction by 1998, and phaseout by 2000

Halons: development of more information on "essential needs" as a basis for the working group to recommend phaseout targets

CT and MC: 50 percent reduction by 1992 or 1993, 85 percent reduction by 1998, and phaseout by 2000

HCFCs: controlled to permit use only in "critical" products as decided by the parties

Although all governments participating in the working group agreed on the general desirability of accelerating the protocol's reduction schedule, the consensus did not hold when it came to the details for specific substances, the timing of cutbacks, and the base year to be used for calculating reductions of newly added regulated chemicals. Also new to the deliberations was the question of how to treat several dozen additional CFCs and halons, most of them theoretical compounds about which little or nothing was known.

Meeting in November 1989, the working group agreed unanimously to eliminate by the year 2000 the five *chlorofluorocarbons* already listed in the protocol, as well as 10 other fully halogenated CFCs. Discussion of possible interim reduction targets, however, brought out differences.

The European Community, joined by Australia, Austria, Canada, Finland, New Zealand, Norway, Sweden, and Switzerland, went beyond the bureau's recommendation and proposed advancing the protocol's target date for the 50 percent reduction to 1991–92 from the existing 1998–99. Most of these countries also endorsed a new interim reduction stage of 85 percent by 1995. The United States and Japan, however, though supporting the 2000 phaseout, wanted more time for the interim reductions. The relative hesitancy of the United States vis-à-vis the EC on this point was attributable to the EC's ability to achieve the 50 percent cutback almost immediately by eliminating CFCs in aerosols. Since the United States had already taken this step in 1978, it needed more time to introduce as-yet-undeveloped alternatives for other uses.

With respect to the *halons,* the scientists had developed a new list of 46 substances in addition to the 3 already controlled under the Montreal Protocol; nearly all of these, however, were theoretical compounds that had not actually been produced, much less used. They were designated in the working group deliberations as "other halons," and, although very little was known about their characteristics, some concern was expressed that they should be included in the protocol to forestall their possible development and use as substitutes.

Initially, Australia, Canada, Finland, New Zealand, Norway, Sweden, Switzerland, and the United States advocated a complete phaseout of all the halons by the end of the century. Some suspected that the United States could support an early production ban because the U.S. Department of Defense had stockpiled substantial halon inventories for its fire-protection needs. In contrast, the EC, Japan, and the Soviet Union were greatly concerned about the continuing unavailability of alternatives in some critical uses. At the November 1989 Geneva meeting, the Soviet Union proposed a reduction of only 10 to 50 percent by 2000; the Soviet representative declared, with some emotion, that no known substance could replace halons in protecting from fire the priceless art treasures of Leningrad's Hermitage Museum.

By March 1990 there was a growing consensus for a 50 percent halon reduction by 1995 and a virtual phaseout, with exceptions for essential uses, by 2000. The EC still held to a 2005 phaseout date. But Canada and the Nordic states went even further, now arguing that halons were rapidly losing their irreplaceable character. They proposed to advance the production phaseout to 1995, with use of stored halons to meet any remaining needs up to the year 2000. A subsequent U.S. diplomatic mission to Moscow gained Soviet acquiescence in an end-of-century phaseout, provided that exceptions were allowed for specific essential uses agreed by the parties. This formula also proved acceptable to Japan. In reality, research was by now moving so rapidly on alternative fire-fighting technologies that knowledgeable observers believed there would soon be no applications in which halons would be indispensable.

New Chemicals, Old Controversies

With respect to chemicals not yet covered by the protocol, there was a consensus early in the negotiations that both methyl chloroform and carbon tetrachloride should be added to the list of controlled substances. Several countries favored eliminating *carbon tetrachloride* as early as 1995, but most looked toward 2000. There were various proposals for interim reduction steps of 20, 50, and 85 percent scattered throughout the decade. Although alternatives to this highly toxic chemical were already on the market, it would take some time to ensure their availability and acceptance at the myriad points of production and use, especially throughout the developing countries. It was understood that continuing use of CT as a feedstock in the production of other substances, during which it was chemically transformed and emitted no chlorine, would be allowed.

Methyl chloroform, however, proved a much more contentious issue, chiefly because of the strong position taken by industry. At the first negotiating session in August 1989, the extremes ranged from a Japanese proposal merely to study the problem to a Soviet recommendation to ban MC by 1992; apparently the Soviet Union used little or nothing of this substance. The Nordic countries, together with Australia and New Zealand, were already advocating a phaseout by 2000.

By November the EC, Japan, and the United States all agreed on a freeze by 1991–92. Japan and the EC (under pressure from France and the United Kingdom) were unwilling to commit to more than reassessing the situation at a later time. However, the U.S. delegation proposed reductions of between 25 and 100 percent by the year 2000, depending on the results of future assessments; this wide range also reflected interagency disagreements, with EPA clearly favoring the phaseout. The list of governments strongly endorsing a 2000 phaseout, with interim reductions of 50 percent in 1992–93 and 85 percent in 1998, now expanded to include Australia, Canada, Finland, New Zealand, Norway, Sweden, and Switzerland.

By the March 1990 meeting there was some further movement by the EC and Japan beyond a freeze. The Community proposed reducing methyl chloroform by 25 percent in 1994, while Japan suggested a 20–30 percent cut by 2000, with an eventual phaseout in the middle of the next century. There would clearly be a wide range of positions regarding cutbacks of methyl chloroform at the June meetings.

There was an even greater divergence of opinion within the working group over the *hydrochlorofluorocarbons.* The year in which chlorine concentrations were expected to peak would depend on the timetable for phaseout of the more potent CFCs, CT, and MC and would occur shortly after the turn of the century regardless of any actions taken on HCFCs. The central issue was to set a phaseout date for HCFCs that was not too early to discourage development of these substitutes, but not too late to increase unacceptably the risk to the ozone layer. If HCFCs were limited in the near term, chlorine levels would start sooner on a gradual downward path to the 2 ppb chlorine target in 2070–2080. A later phaseout of HCFCs would involve accepting higher chlorine levels for several decades longer before a steeper, but later, descent to the same point (see Figure 11.2). A related question was how to encourage the development and use of those HCFCs with relatively shorter lifetimes at the expense of those with greater chlorine-loading potential.

At the August 1989 meeting, many participants, most notably the European Community and Japan, opposed any limits at all on this family of substitutes for the stronger ozone-depleting substances. Even the bureau in September could not come up with specific limits or dates. By the November working group session, however, a coalition of smaller industrialized countries was once again at the forefront. Australia, Finland, New Zealand, Norway, Sweden, and Switzerland proposed new protocol language that would restrict most of these transitional substances to specified critical uses for which no alternatives existed, and would phase them out completely between 2010 and 2020. A U.S. proposal would eliminate HCFCs for new equipment in 2020–2040 and ban all production in the period 2035–2060. However, other participants, including the EC, Japan, and the Soviet Union, still opposed including these compounds in the protocol as controlled substances, even for mandatory reporting purposes.

The only major change by the March 1990 meeting was Japan's announcement of support for a phaseout in the middle of the twenty-first century. Everyone agreed that further assessment of the situation was essential. In the face of such adamant resistance by major parties to including HCFCs in the protocol, the proponents of controls began to consider a fallback position of treating them in a separate "resolution" by the Second Meeting of Parties. Though nonbinding, such a statement would at least send a message to industry that ozone-depleting substitutes should not be considered a long-term solution.

The debates over methyl chloroform and HCFCs centered on the degree of risk involved in continuing their use over a longer period. The governmental advocates of stronger regulations—which now had substantial vocal support from such international environmental organizations as Greenpeace and Friends of the Earth—were opposed to keeping atmospheric chlorine concentrations near peak levels for too long. They argued that delay in bringing chlorine down to precrisis levels posed unnecessary risks, particularly as an extra margin of safety might be needed in the event of incomplete compliance with a phaseout of CFCs, CT, and halons. Other participants, however, maintained that there were also risks for the ozone layer in imposing overly restrictive limits on the substitutes, particularly HCFCs; if their development were discouraged, there was a danger that usage of the more-damaging CFCs would be prolonged.

As in 1987, the proponents of tighter controls hoped to force the pace of technological change. They aimed at discouraging both producing and consuming industries from becoming too dependent on ozone-depleting substitutes. Their objective was, through clear signals, to provide a greater incentive for industrial research on a new generation of chemicals and technology that would neither harm the ozone layer nor aggravate greenhouse warming.

In sum, achieving a recovery of the ozone layer was proving to be a far more complex task than it had appeared back in 1987. The working group had identified no fewer than 46 halons in addition to the originally controlled 3, plus 10 more CFCs over the original 5—all now to be considered for regulation. At least 34 HCFCs were listed as "transitional substances." The inclusion of new chemicals, for which ozone-depletion-potential values might not be available or reliable, would further complicate the calculation of the "baskets" of substances for control purposes.[34]

The continuing lack of consensus among governments on MC and the HCFCs also conferred an unanticipated and troubling importance on a hitherto little-noticed aspect of the protocol: the decision at the final 1987 negotiating session that individual parties to the treaty could, in effect, opt out of new controls on substances not on the original list agreed to at Montreal; such new controls would have to be in the form of a protocol amendment, which required formal ratification and was not binding on nonratifying parties.[35] This provision could now lead to a complicated situation in which some parties to the protocol would be bound by controls over the newly added substances while other parties were not. Several intriguing questions arose from this prospect. For example, could parties be obligated to provide data on substances that they did not accept as being controlled under the protocol? Would it be necessary to redefine the concept of "nonparty" according to whether or not a given country was committed to controls over particular substances? How would the treaty's trade sanctions be applied to parties that did not accept controls over all substances?[36]

Although these unexpected twists could lead to late-night sessions for international lawyers in several foreign ministries, the general direction of the evolution of the Montreal Protocol appeared clear: protection of the ozone layer had become, in less than five years since the signing of the Vienna Convention, a major worldwide priority.

In conclusion, the Open-Ended Working Group in its preparatory deliberations had made substantial progress toward stronger controls. The original CFCs and halons would be phased out more rapidly than any of the Montreal negotiators had imagined possible. And new substances would be added to the list as a result of the scientific understanding of the threat to the ozone layer that had been gained since the protocol's signing in September 1987. The only question was how stringent would be the treaty revisions approved in 1990.

The South Claims a Role

12

One of the premises of the Montreal Protocol had been that developing countries would be encouraged to join if they could count on a reasonable expansion in use of CFCs and halons during the 10-year transitional period, after which they would move to newly developed technologies and follow the original reduction schedule. Most observers were thus unprepared for the intensity of concerns subsequently expressed by many developing countries despite this concession. The nations of the South, most of which had been onlookers in the ozone negotiations through 1987 as the rich countries argued over chemicals scarcely used in the developing world, moved to center stage in 1989 and claimed a major role in revising the protocol.

By 1989 the objectives of the developing-country negotiators had undergone significant change. At Montreal their preoccupation, reflected in the negotiations over article 5, was primarily to maintain maximum usage of CFCs for the longest possible grace period. But with industrialized countries now on a fast track toward phaseout rather than a 50 percent reduction, the grace period became almost irrelevant. It would now be in the interest of developing countries (or "article 5 parties," as they were increasingly referred to in working group texts) not to linger too long with CFCs, but rather to move as rapidly as possible to new technologies—and to ensure that help was available to accomplish this.

Developing Countries: An Unfair Burden?

Industrialized nations, with less than 25 percent of the world's population, were consuming an estimated 88 percent of CFCs; their per capita consumption was more than 20 times higher than that of the developing nations. For China, the world's most populous country, the disparity was even greater: its per capita CFC consumption was only about one-fortieth that of the European Community and the United States.[1]

In effect, use of these chemicals had for decades contributed to the well-being of the industrialized countries, but at the same time inadvertently built up a threat to the entire planet in the form of long-lasting CFCs in the stratosphere. Developing-country governments, in statements at the 1989 London and Helsinki conferences, stressed that the problem was not of their making. They sought assurances that their populations would neither be deprived of the benefits of these substances nor have to pay more for equivalent products and technologies. They were increasingly worried that the drive toward rapid phaseout could add new burdens to their economies and adversely affect their standard of living. They argued that additional technical and financial assistance was essential to enable developing nations to contribute to the protocol's objectives.[2]

There are several ways in which developing countries might incur incremental burdens through accepting the treaty obligations. In the short term, as CFCs were phased out they might become more expensive to countries dependent on imports. The replacement chemicals and the products made with them were also expected to be costlier. Those developing countries that were themselves current or prospective producers of CFCs and their related products would face problems of access to new substitute technology and the attendant costs of royalties and licenses. In addition, there would be costs associated with converting existing CFC facilities, including the purchase of new capital equipment and possible premature abandonment of old. Operating costs might also rise, including possibly higher-priced raw materials and retraining of workers. Some developing countries also felt that the protocol unfairly excluded them from potentially lucrative trade in products made with or containing CFCs.

It was very difficult to assess the degree of hardship that might be occasioned by these factors. Near-term upward pressure on CFC prices in producer countries could be offset by the protocol provisions that forced consumption in those countries to decline at a faster pace than production, thereby creating excess capacity (article 2). The economic assessment panel believed that this effect would ensure "reasonable prices" for exports to supply the anticipated small transitional needs of developing countries.[3] Other analysts, however, questioned the validity of this conclusion.

The extent of incremental costs associated with the new substances and replacement technologies was also uncertain. There would be conversion costs up front, but there would be a tendency for costs to fall over time as research outlays were recovered and economies of scale and competitive forces came into play. It was conceivable that some substitutes and technologies could yield offsetting savings or could even turn out to be cheaper and more efficient than CFCs, as had been the case with alternative aerosol sprays. There were no guarantees, however, whether and when this would occur, and developing countries insisted that the imprecise promises of the Montreal Protocol regarding external assistance be given more substance. The *Synthesis Report* confirmed that such assistance would definitely be required.[4]

A relevant consideration in this regard was the extent of damage to the ozone layer that could be caused by future demand for CFCs and halons in the developing nations—whether inside or outside the protocol. If many large developing countries were actually to expand consumption in 10 years from existing insignificant levels to the 0.3 kilogram per capita limit allowed by the treaty, the resultant increases could dwarf cutbacks by the industrialized countries.[5] But it was questionable whether such an enormous growth in demand was attainable, and developing-country governments themselves characterized this prospect as "extremely unrealistic."[6]

China's CFC consumption had been rising by 20 percent annually in recent years, and the number of its refrigerators reportedly increased by over 80 percent in 1988 alone.[7] However, the technology assessment panel determined that growing refrigeration needs in developing countries would consume fewer CFCs than had originally been expected. Although refrigeration, air conditioning, and heat pumps together accounted for 25 percent of global CFC consumption, household refrigeration was only 1 percent, and developing countries accounted for only a small fraction of that. Therefore, the panel estimated that even a 30 percent annual increase in domestic refrigerators in all developing countries by the year 2000 would generate a demand for CFCs equivalent to less than 2 percent of total worldwide 1986 consumption.[8]

The longer-term potential for developing countries to undermine the effectiveness of the Montreal Protocol could not, however, be dismissed. The science panel had calculated that a 15-year lag in phasing out CFCs by countries accounting for 10 to 20 percent of total consumption would not materially affect peak chlorine concentrations, and would result in only a minor delay in attaining the target level of 2 parts per billion. But if 20 percent of 1986 CFC consumption were to continue indefinitely, chlorine loading would never fall below 2 ppb. And even long-term noncompliance by countries consuming as little as 10 percent of the world's 1986 total could delay elimination of the ozone hole until the end of the twenty-second century, with imponderable risks for the planet.[9]

Developing countries accounted for over 75 percent of the world's population, and this proportion was growing. Moreover, CFC technology was inexpensive and uncomplicated; plants could be small in scale and rapidly constructed and could achieve a relatively fast payback. Therefore, some developing nations could be tempted to build their own CFC facilities if the only alternative was to purchase more-expensive substitute technology or products from Europe, Japan, or the United States. The treaty's trade restrictions became irrelevant against the sheer size of potential domestic markets in some countries. Furthermore, to the extent that populous nations stayed outside the protocol and were therefore not bound by the trade restrictions, new producers in developing countries might try to supply CFCs and CFC products to Africa, Asia, and Latin America even as industrialized-country manufacturers were phasing down.

The potential for large-scale nonaccession to the Montreal Protocol or noncompliance with its objectives did not appear entirely academic to some observers. A 1987 Rand Corporation study predicted that the 13 developing countries with the highest demand for CFCs by the year 2000 would be, in order of potential use, China, India, Brazil, Saudi Arabia, South Korea, Indonesia, Nigeria, Mexico, Turkey, Argentina, Venezuela, Algeria, and Iran.[10] When the Open-Ended Working Group began its negotiations in August 1989 to revise the protocol, only 3 of these countries—Mexico, Nigeria, and Venezuela—had ratified. Moreover, only 14 developing nations had become parties to the protocol by that date; an additional 7 had signed but not yet ratified. In contrast, virtually every industrialized nation, large and small, had joined the protocol.

However, governments of developing countries could not regard continued emissions of CFCs and halons with indifference—they also had a stake in protecting the ozone layer. Even though harmful ultraviolet radiation would cause relatively greater incidence of skin cancer among lightly pigmented populations, all people are susceptible to suppression of the immune response system and to eye cataracts. Indeed, poorer general health conditions and medical facilities increase the risks for populations in developing countries from these prospective health threats. Similarly, productivity declines in agriculture and fisheries would have a disproportionate impact on the developing world, where many already subsist at the margin and food shortages are common. In addition, damage to materials, plastics, paints, and buildings from increased ultraviolet radiation would be more severe in the tropics than elsewhere. The physical threats were real enough.[11]

But there was another, less tangible factor that might motivate otherwise hesitant governments to join the international effort to repair the ozone layer. A new wave of ecological consciousness was uniting populations and governments from every region in common concern for protecting the environment. Scientists, political leaders, international organizations, and ordinary citizens were all part of this phenomenon. UNEP itself, led by an Egyptian and the only major UN organization headquartered in a developing country, represented in many ways the aspirations and dignity of the South; as the creator of the Montreal Protocol, UNEP had a strong institutional interest in its success. And the United Nations system could be a powerful source of moral suasion. In short, failing to accept a share of responsibility and opting actively to threaten the ozone layer would not be an easy course for a government to follow.

Negotiating for Aid

The creation of a financial mechanism and the related question of modalities for transfer of technology proved to be the most difficult issue in the entire treaty revision process. This fact did not reflect any lack of good faith among the participating governments. Quite the opposite: there was broad agreement on the desired objectives, and the debates were characterized by a pragmatic and collaborative spirit and a virtual absence of political rhetoric. Developing countries desired some mechanism that would ensure contributions by industrialized countries to cover incremental costs of the phaseout and transfer of replacement technologies. For their part, industrialized-country governments both accepted a sense of responsibility for the situation and recognized that their own efforts to restore the ozone layer would be jeopardized if the developing world could not, or would not, cooperate. Helping developing nations to bypass CFC technology would be a good investment when measured against the potential costs of even greater damage to the ozone layer.

But despite willingness in spirit, the negotiations proved extremely arduous. No other subject required so many meetings and consultations or generated so much documentation. There were many complex facets to the issues, and the governments sensed that they could be establishing precedents with possible important future implications for North-South relations.

The Open-Ended Working Group considered these subjects at four meetings—August and November 1989 and February and May 1990—plus a major portion of its marathon wrap-up session in June. At its very first meeting in August, which was attended by six article 5 parties and twelve noncontracting developing countries, in addition to the industrialized nations, the working group began tentatively to examine what would be involved. Led by Mexico and Venezuela (parties to the protocol) and China and India (nonparties), developing-country representatives outlined four initial basic concepts:[12]

A discrete multilateral trust fund should be established within UNEP to meet all incremental costs to developing countries of complying with the protocol.

The fund should be financed by "legally enforceable obligations" from industrialized countries, on some agreed burden-sharing basis.

Such contributions should be additional to, rather than a diversion from, existing aid flows.

"Free access . . . and non-profit transfer" to developing countries of safe technologies should be guaranteed.

Although industrialized-country delegations acknowledged the equity of developing nations' concerns, it was evident that details of these proposals would require considerable negotiation. A continuing aim of donor governments was to ensure that assistance be used effectively and specifically to advance the protocol's objectives. For this reason, many donors were traditionally opposed to creating new institutions, especially ones that might be outside their control. Japan, the United Kingdom, the United States, and other major donors strongly preferred to channel aid through bilateral programs or existing multilateral institutions such as the World Bank. The suggestion that donor contributions should be mandatory rather than discretionary raised some eyebrows, although there was no objection to establishing a target scale of payments based on some agreed-upon formula. The proposed guarantee of technology transfer also raised thorny issues of intellectual property rights and patents.

A crucial and universally recognized area of uncertainty was how much money would actually be needed. Outside the working group, there had been some talk in developing countries of employing the global environment as a lever for redressing North-South economic inequalities: enormous resource transfers were mooted as the price for enlisting developing nations' cooperation in protecting the ozone layer. To their credit, the Montreal Protocol negotiators avoided such rhetoric.

On the basis of a Netherlands study by a private consulting firm, the UNEP secretariat offered an initial estimate of $400 million annually for 10 years as the incremental cost to developing countries, but several delegations questioned the assumptions underlying the model. There was general agreement that in order to determine probable requirements, more information was needed both on current and future demand for CFCs in the developing world and on options and costs of reducing their use. A U.S. proposal immediately to begin country-specific case studies was accepted by the working group, as was the suggestion to commission an analysis of the possible roles for existing and new financial mechanisms. Finland, the Netherlands, Norway, Sweden, the United Kingdom, and the United States offered to finance the studies.

At its November meeting, the Open-Ended Working Group focused primarily on control measures rather than on aid. However, since this was the first time the group drafted possible amendments to the protocol, the representatives of more than 20 developing countries in attendance—mainly nonparties—took the occasion to express their convictions in treaty language. Led by the resourceful Mexican ambassador Juan Antonio Mateos, the article 5 parties proposed to remove the condition in that article that permitted developing countries to increase production and consumption of the controlled substances only in order to satisfy their "basic domestic needs." Such an action would reverse a decision of the First Meeting of Parties at Helsinki and open a large loophole to expand use of CFCs in developing nations for incorporation into products in international trade. This threat of a substantial worldwide growth in use of CFCs remained in the draft text until the final hours of the London conference; it was dropped in the context of settling the other outstanding developing-nation issues.

The Mexican-led group accorded greater importance to three pregnant clauses that they introduced into the draft amendment text in November. One provided that "the obligation [of article 5 parties] . . . to comply with

. . . the control measures . . . will be subject to the transfer of technologies and financial assistance." A second stated that technology transfer should be made on a "preferential and non-commercial basis." And the third would create an "International Trust Fund . . . established within UNEP," to be managed by a committee with equal representation from article 5 parties and donor countries.[13]

Because the studies relevant to these subjects commissioned in August were just getting under way, the working group did not consider it useful to debate this language at its November meeting. The negotiators made up for this initial restraint, however; these phrases were to become the centerpiece of seemingly endless future deliberations.

At the February 1990 working group meeting, the first efforts were made to reduce uncertainties concerning the size of the financial mechanism and to elaborate some principles for its structure.[14] At Mostafa Tolba's request, the U.S. Environmental Protection Agency provided "very preliminary" estimates of possible incremental costs to developing countries. EPA estimated that for the next three years, such costs could amount to some $100 million for existing article 5 parties, plus an additional $100–200 million should China, India, and others join the protocol in the near future. U.S. officials noted that it was almost impossible to make meaningful financial estimates for more than three years into the future because of the rapidly changing technology and the development of previously unknown chemicals. The United States therefore proposed, and the working group accepted, the idea of a three-year rolling fund with a budget subject to periodic revision.

Tolba proposed to the group that a new multilateral fund be established with mandatory assessed contributions by industrialized countries on a principle of additionality; that is, these contributions would be additional to existing aid flows. Tolba viewed the new fund as a "safety net," supplementing existing bilateral and multilateral aid channels. He proposed that UNEP, as secretariat of the protocol, have a central role in "catalyzing and coordinating" the work of other organizations assisting in the new mechanism. Similar principles were echoed in a proposal submitted by China, Finland, the Netherlands, and all other developing countries as a bloc.

During the ensuing debate, there was considerable resistance to creating a totally new institution. Representatives of major donor countries stressed the difficult and time-consuming process involved. Some donor governments favored only the establishment of a "clearinghouse" to provide objective information to article 5 parties on the availability of aid from existing bilateral and multilateral sources and to facilitate requests for assistance.

There was a growing recognition, however, that something new and different was required, and gradually even the donors began to use the term *multilateral fund.* It was suggested that some structure could be designed, with minimum bureaucracy, that could incorporate the particular expertise of UNEP, the World Bank, the United Nations Development Programme (UNDP), and other relevant agencies. There was widening agreement that such a mechanism should be under the policy control of the parties to the protocol, rather than, say, the board of directors of some other organization. Tolba and many developing countries clearly hoped to maximize UNEP's own influence in the fund, while the major donor nations preferred a stronger role for the World Bank, with its extensive financial and development experience. This question, therefore, remained open.

The official report of the February working group meeting stated that "all delegations approved the principle of additionality."[15] Agreement on this principle appeared to meet one of the most important demands of the developing countries concerning the proposed new fund. In reality, however, the statement concealed a continuing reservation by the United States. Throughout the negotiations, the U.S. delegation had consistently maintained that it viewed the term *additionality* as implying the allocation of more funds specifically to ozone protection, but not necessarily the allocation of funds in addition to overall foreign assistance flows. In this sense, it joined the consensus, while indicating that a more generous interpretation of the term would require approval at higher levels in the U.S. government.

The February meeting made progress in addressing several concerns of potential donors. Agreement was reached that funding assistance could be provided in the form of both concessional loans and grants rather than grants only. The group also tightened the definition of incremental costs eligible for financing: specific categories of costs were proposed, and any estimate of needs would have to reflect the most cost-effective options, take account of potential savings incurred in replacing CFCs, and avoid double counting of costs. In another concession to the donors, the clearinghouse function was incorporated into the emerging funding mechanism to assist countries in identifying needs, to facilitate other bilateral and multilateral aid, and to underwrite technical cooperation and training; about a dozen country studies, including ones for Brazil, China, India, and Mexico, were already in process. The critical issue of whether contributions to the new fund would be obligatory or voluntary remained, however, unsettled.

The Drive for Guaranteed Technology

For a number of newly industrializing developing countries, technology transfer was a separate issue from financial aid. More was involved than simply subsidizing developing-country purchases of CFC substitutes, or even establishing on their soil affiliates of foreign companies utilizing the most modern technologies. For countries such as Brazil, China, India, and Mexico, there was a matter of principle involved: they could already produce CFCs on their own—therefore, they also wanted to be able to produce any new substance on their own, without being subject to potential exploitation by large foreign patent holders. It was a question of *guaranteed access* to new technologies, on terms they could afford.

However, the idea of "preferential and non-commercial" transfer of technology, as introduced by the Mexican-led group at the November negotiating session, posed a challenge and a dilemma to the industrialized world. Entrepreneurs would be reluctant to invest sizable sums in research and development of new products, only to have their results given away to companies in developing countries that could then undercut potential markets. There was a principle involved here also: the private sector had to be able to recoup its costs in order to maintain an incentive for investment in technology development. In addition, it was puzzling for Western officials to contemplate how governments could enforce transfer of private-sector technology on noncommercial terms. The World Intellectual Property Organization agreed that in such matters governments have little influence with private companies beyond persuasion and incentives.[16]

For its part, industry regarded with some alarm the attempt to legislate technology transfer by treaty. Its view was that information exchange is most effective when it occurs directly between industry representatives and not through government channels. The International Chamber of Commerce (ICC) felt that it was in any case premature to transfer technology for potential CFC alternatives until they had been thoroughly tested for safety and had demonstrated real market potential. At such time, the most likely modalities for transfer of ozone-safe technologies—as for other technologies in the past—would be joint ventures and licensing arrangements. And such investment would gravitate, via the market mechanism, toward those developing countries where the investment climate was favorable, market prospects good, and intellectual property rights respected.[17]

The ICC stressed that the forces of international competition would militate against unfair bargains. The fact that at least 10 companies were racing to develop alternative refrigerants should assure developing countries of access and choice at reasonable cost. In addition, in order to stimulate demand for their new products, producers in the industrialized countries would be likely to provide information on related "downstream" technologies to user industries in developing nations.

In the meantime, even before substitutes became available, considerable technology transfer was in fact taking place to enable article 5 parties to reduce their dependence on CFCs in the short term, and there was scope for more. Current priorities for such transfer included recycling, reclamation, and conservation technologies and replacing CFC usage with already available methods in aerosols and aqueous solvents. Consortia of U.S. refrigeration, air conditioning, and electronics firms, in cooperation with EPA, were planning to make nonproprietary information available to developing countries, following the example of Digital Equipment Corporation in electronics cleaning technology. Industry representatives discussed with UNEP their plans to utilize technical guidebooks, electronic databases, and training workshops to transfer new information to developing nations.[18]

It was clear that resolving the technology transfer issue was essential to bringing the major developing countries under the protocol's regime—and that doing so would require further creative thinking and new forms of cooperation among industry, governments, and international agencies.

A Matter of Additionality

During the weeks leading up to the working group meeting on these subjects in May, Tolba held informal consultations on funding and transfer of technology with small groups of government representatives and with the World Bank, UNDP, the United Nations Industrial Development Organization, the World Intellectual Property Organization, and the ICC. The

World Bank expressed strong interest in participation, regarding ozone protection as a major element in a new billion-dollar "Global Environmental Facility." Bank President Barber Conable wrote Tolba in April that this "pilot facility . . . could demonstrate a new form of collaboration between UNEP, UNDP and the World Bank."[19]

The issue of aid additionality, however, which had been papered over at the February meeting, was about to reemerge. On the eve of the May 9 working group session, the *Washington Post* learned that a new U.S. position would be announced in Geneva, supporting a mechanism within the World Bank for ozone protection provided that funding came from existing bank resources and that no additional donor contributions would be required. The *Post* reported that the decision had been made by White House Chief of Staff John Sununu and Budget Director Richard Darman, overruling EPA Administrator William Reilly.[20]

The U.S. position reflected a belief, not in itself unreasonable, that a reassessment of projects and priorities in the multibillion-dollar World Bank lending programs could well come up with the relatively modest sums initially required for an ozone fund. The U.S. administration was greatly concerned that acquiescence in the proposed financial mechanism for ozone would create a precedent for a future global authority to deal with the vastly more expensive problem of greenhouse warming. Tolba and others had in fact made just this point of precedence during discussions over the previous months.[21] Several European and other governments had already endorsed a global climate fund in principle, and the U.S. administration was uneasy about the possibility of unpredictably large future demands to aid developing countries in reducing greenhouse gas emissions and adapting to the consequences of climate change. At least one commentator noted a philosophical consistency between the new U.S. policy and a speech by Darman a week earlier at Harvard University, in which the budget director had been critical of a type of "anti-growth, command-and-control, centralistic environmentalism" that was aiming at "global management."[22]

Whatever the rationale, the U.S. position was widely interpreted as a policy reversal that could upset the progress made toward strengthening the Montreal Protocol and attracting more developing country participation. Domestic reaction was immediate and critical. *Time* headlined "A Baffling Ozone Policy," and the *Washington Post,* even while expressing support for the administration's cautious approach to global warming, editorially labeled the new ozone policy "a serious mistake."[23] It appeared to many that the United States had abandoned a commitment to the consensus on aid additionality that had been achieved by the February working group—even though technically the U.S. delegation had at that time only indicated an intention to raise the interpretation of the additionality concept at higher levels. Nonetheless, doubts were expressed about the reliability of the United States as a negotiating partner.

Many observers expressed astonishment that the United States would balk at pledging relatively small sums of $8.3–25 million annually for a three-year period, when these costs were weighed against the benefits of engaging the developing world in the global effort to protect the ozone layer. American industry offered strong public support for the proposed ozone fund. Industry spokesmen pointedly observed that the U.S. contribution was minimal in comparison with an estimated $5.7 billion that the federal government would collect over the next six years from a newly imposed excise tax on CFCs and halons, designed to encourage their replacement and capture windfall profits. Privately, several corporate leaders, including the chairman of Du Pont, informed the White House of their firm backing for the fund.[24]

In separate letters to President George Bush, 24 Republican and Democratic senators stated their concern that diplomatic isolation of the United States on this issue would threaten loss of its position as a world environmental leader. The 12 Republican senators called the position "penny wise and pound foolish." They added that any worries over setting precedents were misplaced, since a global climate fund, if it did eventuate, "can and should be debated on its own merits."[25] The senators asked the president to reverse the policy, and legislation was rapidly introduced in both houses of Congress to provide funding for a U.S. contribution in case the administration maintained its position.[26]

The response from delegations at the May working group meeting in Geneva was equally swift and uniform. Representatives of industrialized and developing nations alike expressed "concern and disappointment" and "deep dismay."[27] An official of the World Bank declared that it would participate only if additional funding was made available. All major donors, including the EC and Japan, reaffirmed their commitment to additionality. Several speakers expressed fear that the success of the treaty had been jeopardized, and China and India made clear that they would not accede under these circumstances. There was a general call for the United States to reconsider its position before the London conference.

The furor over the new U.S. policy was especially ironic because many innovative elements of the emerging financial mechanism had originated in U.S. proposals during the intergovernmental deliberations. Furthermore, EPA was playing a major role in contributing funding and technical expertise to the country feasibility studies that would define the scope of the needs. The administration had evidently misgauged the intensity of international feeling over this issue: developing countries as well as other donor governments all regarded aid additionality in the case of the ozone layer as a matter of equity. It was also ironic that, had the U.S. statement not specifically focused attention on additionality, no one could ever demonstrate, among the billions of dollars of aid flows fluctuating from year to year, whether any given program in a given country represented additional funding or not. But the semantics were, for sensitive participants in the process, a critical signal.

The working group, after absorbing the shock on the additionality issue, turned in a constructive spirit to other aspects of a multilateral fund, as now incorporated in a document offered by Executive Director Tolba. The U.S. delegation was able to contribute to further progress in developing the new fund. Agreement was reached on the list of incremental costs eligible for support, which would now be appended to the fund's terms of reference. Developing countries for the first time accepted the concept of a tripartite division of responsibilities among the World Bank, UNDP, and UNEP. It was also agreed that a donor's bilateral assistance could be counted, up to a percentage still to be decided, as part of its contributions to the multilateral fund. On the issue of mandatory versus voluntary contributions Tolba attempted to bridge the gap between developing countries and major donors by promoting the concept of "voluntary contributions on an assessed basis," which conveyed a sense of implicit obligation.[28] The still-undecided options for determining a donor's target contribution were the traditional UN scale of assessment or the donor's 1986 consumption of CFCs.

* * *

With this meeting, the Open-Ended Working Group completed almost nine months of negotiations. Considerable progress had been made, under UNEP's stewardship and with the imaginative leadership of the seemingly indefatigable Dr. Tolba, in refining issues involving more stringent control measures and developing-nation participation in the Montreal Protocol. More and more countries were engaging in the process. The scientific and academic communities were contributing their expertise. The issues were being debated in a constructive spirit, and an informed public—now including environmental groups as well as industry associations—was following the procedure closely.

But many crucial items remained unresolved. The working group would meet for one last time in London in June, immediately before the Second Meeting of Parties convened to approve revisions to the treaty. Much important work was still ahead.

Strong Decisions in London

13

In June 1990, delegations from governments, international institutions, and private-sector organizations converged on London to consider and decide upon significant revisions of the 1987 Montreal Protocol on Substances That Deplete the Ozone Layer. By that time, 58 governments plus the European Community, representing 99 percent of estimated world production and 90 percent of consumption, had ratified or acceded to the protocol. Thirty of the parties were industrialized countries; 28 were developing countries; a notable recent addition was Brazil. Attention was riveted on China and India, which had not yet joined.

On the Eve

A month before the June 20 scheduled beginning of the final session of the Open-Ended Working Group, Executive Director Mostafa Tolba circulated for consideration a "personal" proposal for revisions of the protocol's control measures.[1] This four-page document built upon the deliberations of the March working group meeting (discussed in Chapter 11) as well as on subsequent informal consultations with officials of representative governments. Tolba offered the proposal in an attempt to promote greater consensus on the measures. The major elements, which formed the basis for the London negotiations, were as follows:

1. *Chlorofluorocarbons (CFCs):* freeze beginning in mid-1989 (already in the protocol); a six-month advance of the 20 percent reduction to begin January 1, 1993; 85 percent reduction in 1997; phaseout in 2000.
2. *New CFCs:* as above, without the freeze stage.
3. *Halons:* freeze in 1992 (already in the protocol); 50 percent reduction

in 1995; phaseout in 2000; parties to decide in 1992, with subsequent review, whether any identified "essential uses" should be exempted from this schedule because of unavailability of substitutes.

4. *"Other halons":* no firm phaseout schedule (because of uncertainty about the properties of these compounds); a (nonbinding) resolution requesting reports on production and use and calling on parties to refrain from using currently unregulated halons except as transitional replacements in essential applications.
5. *Carbon tetrachloride (CT):* 85 percent reduction in 1995; phaseout in 2000.
6. *Methyl chloroform (MC):* freeze in 1993; 30 percent reduction in 1995; 50 percent reduction in 2000; plus a (nonbinding) resolution proposing phaseout not later than 2010, subject to future reviews.
7. *Hydrochlorofluorocarbons (HCFCs):* mandatory reporting of production, exports, and imports; phaseout not later than 2040; resolution calling for phaseout, "if possible," by 2020.

Tolba's compromise package attempted to reach a balance among the many contending proposals that were embedded in the heavily bracketed text produced by the working group in March after months of deliberations. During the period since Helsinki, however, several governments had moved forward on their own with more stringent measures. By June 1990, Australia, Austria, the Federal Republic of Germany, the Netherlands, New Zealand, Norway, Sweden, and Switzerland had independently announced planned cuts in CFCs of 90–100 percent as early as 1995.

Meeting in early June (after Tolba's proposal had been circulated), the EC Council of Ministers determined that a 50 percent reduction in CFCs by 1991–92, 85 percent by 1995–96, and 100 percent by 1997 "or at the latest before 2000" would be "reasonable and realistic" objectives for the London conference.[2] Although this seemed to be more stringent than Tolba's compromise, the document concealed differences within the Community, especially with respect to the ambiguity of the phaseout date. As at Montreal in 1987, Belgium, Denmark, the Federal Republic of Germany, and the Netherlands, this time joined by the Commission itself, favored the strongest options, not only on CFCs but also on other substances.

Symbolic of these continuing differences was the action of the Federal Republic of Germany, one of the world's largest producer countries and the biggest in the EC. Having concluded from an internal analysis that alternatives to the major ozone-depleting substances were close to commercial feasibility in nearly all applications, Germany announced in May 1990 that it would phase out CFCs in 1995, halons in 1996, and CT and MC in 1992. Environment minister Klaus Töpfer declared that this action was intended as an example to other countries in the EC and elsewhere that early phaseout of the ozone-destroying substances was possible.[3]

As the June conference approached, the chemical industry appeared focused on securing its past and potential investments in CFC substitutes, especially HCFCs. On the eve of the meeting, both Du Pont and Imperial Chemical Industries (ICI) announced major new production facilities for alternative refrigerants.[4] However, the immediate plans involved hydrofluorocarbons (HFCs), which did not fall under the jurisdiction of the Montreal Protocol. It seemed that industry was awaiting assurances from the London meetings that large-scale investments in HCFCs would not face very early phaseout dates in the revised protocol.

Du Pont declared that it had already spent nearly $250 million in research and development, and that it envisaged a $1 billion program to bring alternative refrigerants to market. Four planned "world-scale" facilities in the United States, the Netherlands, and Japan were expected to begin operations in 1992–1995. Du Pont anticipated that this move would encourage the automotive, appliance, and other user industries to invest in the changes needed to enable them to utilize the replacement chemicals.

For their part, environmental organizations were demonstrating more sophistication than had been the case during the process leading up to Montreal. Both before and during the London meetings, Friends of the Earth International, Greenpeace International, and the Natural Resources Defense Council (NRDC) held press conferences and circulated brochures and briefing sheets to the public, the media, and officials to match the customary public relations output of industry.[5]

The environmental groups were critical of the Tolba compromise for not going far enough. They argued that if Germany could phase out CFCs by 1995, why could other industrialized countries not do likewise? They particularly criticized industry's defense of methyl chloroform and the proposed 50 percent MC reduction target, noting that the parties' own assessment panel had determined a year before that "at least 90 percent" MC reductions by 2000 were technically feasible.[6] The environmentalists also expressed concern that the parties would compromise on measures that would send an inadequate signal to industry regarding the necessity for only limited transitional reliance on HCFCs.

This degree of involvement by the environmental community was a new element in the ozone negotiations, and there were some differences in style among the organizations. Greenpeace was the most confrontational, staging a theatrical "happening" with costumed demonstrators on the Thames River embankment outside the conference hall, and labeling the revised protocol a "failure" even before the meeting started. Greenpeace demanded immediate bans of all chemicals, with no mention of costs, alternatives, or trade-offs. It also, however, focused needed attention on problems of monitoring compliance and industry secretiveness about its data.

Friends of the Earth was more analytical, basing its sophisticated presentations and briefing material largely on the chlorine-loading models employed by the UNEP science panel. To most government participants, however, who had to assess risks pragmatically, some of its positions, such as a 1992 phaseout of CFCs, seemed extreme and unrealistic.

NRDC (which had originated the court case against EPA during the earlier negotiations, cited in Chapter 3) employed a rather subtle approach in London: it combined the most liberal positions espoused during the negotiations by any government on any individual chemical into one optimal, and quite defensible, package of controls—which it proceeded to urge on delegations.

In general, all the environmental groups distrusted industry's motivations and sense of environmental responsibility in the absence of strict regulation. They had no difficulty in reciting examples of misleading industry statements and questionable activities. Such samples as ICI's straightfaced assertion, as late as April 1990, that "the effect of CFCs on the ozone layer is by no means easy to understand . . . nor is the connection entirely proven," did not encourage confidence in industry's commitment to change.[7] Friends of the Earth claimed in congressional testimony to have evidence that Du Pont and other chemical companies were introducing aerosols propelled by MC and HCFCs for hair sprays and deodorants, even though non-ozone-depleting alternatives for such products had long been on the market.[8] ICI's brochures for the London meetings, though nominally accepting phaseout of CFCs and halons, were full of reasons for delay. ICI saw no need to phase out methyl chloroform until "towards the middle of the next century." And it described carbon tetrachloride (which West Germany was phasing out in 1992) as "invaluable" and recommended deferring even a decision concerning possible phase-out until the 1994 meeting of parties.[9]

The debates had a familiar ring. There clearly remained differences of opinion over the acceptable degree of risk to the ozone layer. That industry needed the spur of regulation seemed to most observers obvious. But the issues of availability, safety, and effectiveness of replacement products and technologies could not be blithely dismissed. Excessively forcing the pace could lead to unforeseen or unintended consequences if substitutes did not reach the market in time. There were risks of noncompliance or offshore production that could delay elimination of the problem chemicals and thereby cause more harm to the ozone layer than would have resulted from a less demanding, more pragmatic, ostensibly "weaker" set of controls. It was now up to the negotiators to find a balance.

A Heavy Agenda

The last session of the Open-Ended Working Group began on June 20, preceded by several days of intensive informal consultations organized by UNEP Executive Director Tolba. The group had six days in which to attempt to complete texts for decisions at the Second Meeting of Parties to the Montreal Protocol, which would take place June 27–29.

The task was more formidable and more complex than the one that had confronted the negotiators three years earlier—even if the sense of history had perhaps been greater at Montreal. More than 80 governments and about 30 nongovernmental organizations were now represented in the working group; still more would attend the meeting of parties. There were new agendas and new actors, representing finance, aid, planning, and development ministries as well as environment and foreign affairs. Because of the many variables, alliances among countries shifted depending on the issues, and different delegations assumed leading roles on different subjects.

Nearly 200 bracketed portions of text in four separate sets of documents required resolution for approval at the meeting of parties. Each document was governed by different voting procedures in the event that consensus could not be reached.

First, the parties would have to agree formally on a number of "Decisions," which would include the three following items as well as individual issues ranging from noncompliance to data reporting and procedural matters. Second, the crucial "Adjustments" section comprised the changes in the reduction schedules of already controlled CFCs and halons, which would be binding on all parties.

Third, the equally critical "Amendment" covered all other modifications to the protocol; this long text included the new controls for carbon tetrachloride and methyl chloroform, the establishment of the financial mechanism, and many changes in articles covering technology transfer, obligations of developing countries, and trade restrictions. The amendment would have to be formally ratified and would enter into force only after 20 governments had done so, with a target date of January 1, 1992.[10] The components of the amendment would be binding only on those parties that accepted it. To avoid the unmanageable situation of creating multiple categories of parties to the protocol, the working group had decided to present a single comprehensive package of all proposed changes in a single amendment: a party would have to accept all or none.

The fourth and last document for decision by the parties would be a nonbinding but significant "Resolution" on transitional substances (HCFCs and "other halons").

The treaty was becoming so complex that one of the decisions of the parties was to commission UNEP to produce a "Montreal Protocol Handbook," setting out all amendments, adjustments, decisions, and other material relevant to its implementation.

Major unresolved issues on the eve of the London meetings included:

1. The timing and extent of cutbacks for CFCs, halons, carbon tetrachloride, and methyl chloroform, and the treatment of "other halons" and HCFCs
2. Details of the new financial mechanism, including means of determining policy, voting procedures, the respective roles of UNEP and the World Bank, and the burden-sharing formula for donors
3. Defining a relationship between the obligation of article 5 parties to comply with control measures and the provision of financial aid and access to technology, including the critical question of how to determine whether the assistance provided to a given party would be adequate to enable its implementation of the controls
4. Revised voting procedures for future changes to the protocol, reflecting the developing countries' insistence on parity
5. Noncompliance procedures, including the status of any parties not accepting the new amendment

Because of the size and complexity of the meeting, the organization of the working group's activities was considerably more formal than at Montreal. Two major open-ended subgroups were created, one dealing with financial mechanism and technology transfer issues, chaired by Finnish Ambassador Ilkka Ristimaki, and the second covering the control measures and everything else, chaired by Victor Buxton of Canada, one of the few participants whose experience dated back to the Vienna Convention negotiations. Each of these groups was attended by nearly every delegation. In addition, there were legal, trade, and other smaller technical working groups.

Delegates were relieved at the outset by an announcement on June 15, by White House Chief of Staff John Sununu, of a revised U.S. policy on the funding mechanism, specifically labeled as a proposal by President Bush.[11] Prime Minister Thatcher, and possibly other foreign leaders, had appealed directly to the president to reconsider the May policy in order to avert failure of the London conference.[12] Central elements of the new U.S. position were acceptance of the aid additionality concept coupled with explicit reference to the nonprecedential nature of the new fund.

Despite all plans, goodwill, and heroic late-night efforts, the Open-Ended Working Group was unable to complete its work by the time Prime Minister Thatcher opened the Second Meeting of Parties on June 27. There were now 95 governments and more than 40 nongovernmental organizations present. Evidencing the importance accorded this meeting, nearly all the parties were represented at ministerial or equivalent level. In addition, 42 nonparties attended, most of them developing nations, and many also at ministerial level. Although there were environmental, educational, and church groups in attendance, the nongovernmental representatives were predominantly from private industry. Their composition reflected the breadth of economic interests involved—not only the chemical sector but also the electronics, automotive, refrigeration, solvent, pharmaceutical, appliance, aerosol, and fiber industries.

There were surprises ahead in London. Some of the compromises reached during the working group negotiations did not hold when the ministers arrived, new alliances emerged, and new deals were struck. Large plenary sessions, involving full governmental delegations and observers from private-sector and multilateral organizations, were almost nonexistent. In an extraordinary departure from customary practice, the heads of 95 national delegations—including those from noncontracting governments—met together for hours of secret debates with no supporting staff save for a translator if the principal needed help in a language other than the six (Arabic, Chinese, English, French, Russian, Spanish) for which the UN provided translation. As at Montreal, but on a grander scale, most serious business was conducted in small informal consultations, "contact groups," and constellations, with floating participants revolving mainly around Mostafa Tolba.

At Montreal, however, settlement was reached before the second night of the plenipotentiaries' meeting, and the third day was devoted to ceremony; at London the outcome remained truly in doubt until the third night, and many ministers had to depart hurriedly without the customary congratulatory speeches. Mostafa Tolba, veteran of decades of hard international bargaining, was later to declare that he had "never been engaged in such difficult and complicated negotiations."[13]

The End of CFCs

As the negotiators in London considered the proposed revisions of the control measures, they were influenced not only by the more stringent regulations already planned by a number of governments, but also by some new scientific data presented by Norwegian scientist Ivar Isaksen and by G. O. P. Obasi, secretary-general of the World Meteorological Organization. Although the new data were still provisional, ground-based measurements from northern Europe and Canada appeared to indicate a "very pronounced decline" in ozone concentrations of nearly 0.5 percent annually over a 20-year period, while satellite measurements showed a 2 to 3 percent drop during the last decade over equatorial regions.[14] Such rates of decline far exceeded the model predictions.

Japan, the Soviet Union, and the United States were the strongest advocates in the working group for Tolba's compromise package. Australia, New Zealand, and the Nordic nations, often joined by Austria, Canada, and Switzerland, were the most consistent proponents of tighter regulation and earlier phaseouts. It was probably no accident that most of these countries are relatively close to one or the other polar region; the implications of the Antarctic ozone hole and the prospects for an analogous loss over the Arctic would not have escaped them. Developing-country representatives were generally bystanders to the debates over controls among the larger and smaller industrialized nations.

On the chlorofluorocarbons, Australia, New Zealand, Norway, and the

EC Commission proposed at the working group meeting to raise the 20 percent cut in 1993 to 50 percent, and to advance the 85 percent reduction from 1997 to 1995 or 1996. The same three nations, joined by Austria, endorsed a 1997 phaseout, while the EC representative artfully suggested elimination "before 2000." Some observers noted that enthusiasm for the earlier phaseout was not uniform throughout the Community, but that the Commission was enjoying the opportunity to turn the tables from Montreal and take an ostensibly stronger position on CFCs than the United States.

Flamboyant EC rhetoric in the working group on the importance for the ozone layer of a 50 percent cutback in 1993 also fell into this category, as it was well known that the Community could attain this target by embracing the aerosol ban that the United States had imposed over a decade previously. As an insurance policy, American negotiators placed a block on two proposed protocol changes that were of importance solely to the EC, one dealing with industrial rationalization and the other with aggregated reporting of trade data, discussed below. These clauses, which some London participants jokingly referred to as "hostages," were released by the U.S. negotiators only as part of the final resolution of the disagreements over controls.

When the ministers arrived, many speakers advocated the 1997 CFC phaseout year, including the representatives of Australia, Austria, Canada, New Zealand, Norway, Sweden, and Switzerland, plus Ireland, speaking for the European Community; EC members Belgium, Denmark, the Federal Republic of Germany, and the Netherlands spoke separately, underscoring their individual views. Japan, the Soviet Union, and the United States strongly supported phaseout in 2000.

The main argument for an earlier phaseout was that it would reduce both the peak chlorine loading and the time required to eliminate the ozone hole. Models indicated that, for every earlier year of phaseout, peak chlorine would be about 0.1 part per billion lower, and the 2 ppb level could be crossed 3.6 years sooner.[15] However, these models did not take account of interim CFC reductions beyond a freeze, which could substantially influence any comparison of phaseout dates. As a result, some observers mistakenly exaggerated the effect of the three-year difference between a phaseout in 1997 and one in 2000; they overlooked the strongly mitigating impact of the interim reductions, particularly the 85 percent cutback in 1997.

In addition, given the current state of technology, Du Pont and other producers were still very doubtful that a total phaseout could be achieved by 1997. For the United States, the difference could be important: it had a much larger existing capital stock (air-conditioned buildings, automobiles, and so on) than the Europeans, which even with improved conservation and recycling might require some continued servicing with CFCs—and it was not yet clear if sufficient alternatives could be on the market by 1997. Even UNEP's technology assessment panel had not concluded that a complete phaseout before 2000 was technically feasible.[16] In the end, this factor proved decisive.

The final compromise reached by the ministers on the CFC schedule comprised the following elements: (1) the 20 percent cutback targeted in the Montreal Protocol for 1993 was dropped as no longer relevant, since virtually all industrialized countries had already passed this milestone, and a 50 percent reduction by 1995 was introduced in its place; (2) the 85 percent reduction proposed by Tolba remained for 1997; and (3) the phaseout in 2000 was confirmed. (For all London revisions see Appendix C; for CFC controls, see annex I, article 2A in that appendix.)

One necessary and interesting bit of housekeeping on CFCs was an ingenious proposal by Tolba, greeted with universal relief, to eliminate the awkward July 1–June 30 control year from the original protocol, with its correspondingly difficult reporting requirements. During the 18 months following the London revisions (July 1, 1991, to December 31, 1992), CFCs would be frozen at 150 percent of the 1986 base year. Thereafter, all controls for all substances would apply to normal calendar years.

At the close of the conference, 13 heads of delegation issued a formal declaration stating that they were "convinced of the availability of . . . alternative[s]" to CFCs and that there was a "need to further tighten" the revision just approved. Australia, Austria, Belgium, Canada, Denmark, the Federal Republic of Germany, Finland, Liechtenstein, the Netherlands, New Zealand, Norway, Sweden, and Switzerland committed themselves to phase out CFCs "as soon as possible but not later than 1997" (see Appendix C). Despite the European Community's rhetorical support of the 1997 phaseout during the London negotiations, the EC Commission did not sign this declaration, nor did 8 of the 12 EC member countries, including the major producers France, Italy, and the United Kingdom.

Most observers regarded the London outcome on CFCs as a strong and realistic compromise. The introduction of the 50 percent cut (in actuality, a three-and-a-half-year advancement of this stage from the original protocol schedule) was both meaningful and unexpected. In terms of the effect on chlorine loading, the sizable interim reductions of 50 and 85 percent reduced the difference between a total phaseout in 2000 and one in 1997 to near insignificance. Moreover, a new clause committed the parties in 1992 to "review the situation with the objective of accelerating the reduction schedule."

The intent was very plain: this was not the last word on CFCs, and the parties would act again on the basis of updated assessments of the rapidly evolving technology. A key member of the science panel, who was present in London, expressed strong satisfaction with this outcome.

Further Phaseouts

Control measures for other chemicals were also subject to hard bargaining. Major controversy over *methyl chloroform* persisted at the London meetings, with a surprising outcome.

During the working group session, a new alliance of the EC, Japan, and the United States supported the Tolba proposal for a 1993 freeze, followed by reductions of 30 percent in 1995 and 50 percent in 2000. The customary coalition of smaller industrialized countries, this time joined by the Soviet Union and West Germany (breaking ranks with the EC), pressed for an 85 percent cutback by 2000.

The key to the ministers' resolution of this impasse was once again the findings of the special assessment panels, reinforced by some sophisticated bargaining. Despite the presence in London of numerous representatives of international solvents industries and their claim that stringent controls would create "astronomical" costs, the case for MC began to appear to more governments increasingly indefensible.[17] It was clear from the science panel's findings that an MC phaseout would be the single most important short-run contribution to lowering stratospheric ozone concentrations, quickly resulting in a 0.5 ppb reduction of chlorine loading. And the technology panel had concluded that a 90–95 percent MC reduction by 2000 was indeed technically feasible.[18]

The U.S. position on MC at London was further softened by actions on the home front. Both houses of Congress had recently overwhelmingly approved amendments to the Clean Air Act that included MC phaseout dates of 2000 (Senate) or 2005 (House of Representatives).[19] Industry had underestimated the likelihood of passage of such strong regulations, anticipating that the possibility of a weak international agreement would cause Congress to retreat. Now American industry, as a consequence of its success in influencing a weak U.S. negotiating position on MC, faced the unpleasant prospect that the international treaty would offer its competitors less stringent controls than the domestic regulations.

As the U.S. delegation reconsidered its initial position in the closing hours of the last day in London, Norway and the smaller countries applied additional pressure by withholding approval of certain preambular paragraphs concerning the nonprecedential nature of the new ozone fund that were of particular importance to the U.S. administration (discussed below). That factor, plus the congressional action and the compelling arguments derived from the assessment panels, won over the U.S. delegation to the side of those pressing for strong controls.

The remaining major holdouts to fixing an early MC phaseout date were now Japan and the United Kingdom (preventing EC agreement). West Germany was lobbying hard in the EC for the phaseout. The United States now used a bargaining chip to secure EC acquiescence: it ceased to oppose an EC amendment that would allow Community members to transfer production facilities for controlled substances within the new European single market. With the EC now accepting larger reductions, Japan, despite its multitude of solvents industry onlookers, would not block consensus.

Thus eventuated a major surprise at London, going beyond even Dr. Tolba's dreams: firm agreement to phase out methyl chloroform. It was a case study of how relatively smaller nations—with a strong assist from the U.S. Congress—could prevail over large producer countries.

The final result was a freeze of MC in 1993, reductions of 30 percent in 1995 and 70 percent in 2000, and phaseout in 2005—plus a commitment to review, no later than 1992, the feasibility of even earlier reductions and phaseout (annex II, article 2E; and annex VII, section III). Moreover, future adjustments of the reduction schedule would be binding on all parties.

The *hydrochlorofluorocarbons* had been included for control in the draft protocol amendment produced at the March working group meeting, albeit entirely in brackets because of EC opposition. The Nordic states had proposed language that would limit HCFCs to specific, essential uses agreed upon by the parties, and would phase them out by 2010 or 2020. Subsequently, in its amendments to the Clean Air Act the U.S. Congress had established 2015–2020 freeze dates and 2030–2035 phaseouts for HCFCs. Tolba, with U.S. support, attempted in his London proposal to retain at least a 2040 phaseout date in the binding part of the protocol revisions. But the European Community was unwilling to compromise on this point.

As a result, the only mandatory requirement on HCFCs that was included in the revised protocol was an obligation to report on production, plus imports and exports to parties and nonparties, the same as for other controlled substances (annex II, article 7).

In addition, a nonbinding resolution requested that these "transitional substances" be used only where "more environmentally suitable" alternatives are not available, and that they "be selected in a manner that minimizes ozone depletion"—a weak attempt to skew usage toward the shorter-lived HCFCs. The resolution also provided for regular review of the situation, with a current objective of phasing them out "no later than 2040 and, if possible, no later than 2020" (annex VII, section II). Finally, the parties asked the technology panel to evaluate the need for transitional substances in specific applications; this was all that remained of the original Nordic proposal that use of HCFCs be permitted only for essential purposes as decided by the parties.

The outcome on HCFCs occasioned the strongest criticism of any actions taken at London. The central issue was whether a nonbinding resolution was too weak a signal to influence industry voluntarily to refrain from excessively expanding reliance on HCFCs. Some environmental groups expressed serious disappointment over the failure to impose mandatory controls.[20] But other observers, including Mostafa Tolba and most participating governments, felt that it was a satisfactory result, for the time being.

It was true that industry could now invest with more confidence in the substitutes for CFCs—but only up to a point. Adequate supplies of HCFCs were in fact desirable in the medium term, in order to hasten the exit from the much more destructive CFCs. But the pressure was clearly on to minimize the duration of use of HCFCs, and in such a climate there would be added risks for companies that did not exercise prudence in planning long-term investments. It was likely that industry would take the signal seriously, recognizing that signals in the past—for example, the 1989 Helsinki Declaration—had been translated into stronger controls.

One flaw in the resolution was its inadequate distinction between short- and long-lived HCFCs: it could have more explicitly encouraged the former and discouraged the latter. However, the entire HCFC subject would be revisited by the parties in 1992; most new HCFC factories would not even be built by then.

With respect to the *halons,* the ten *new CFCs, carbon tetrachloride,* and the *"other halons,"* the meeting of parties accepted Tolba's proposals (annex I, article 2B; annex II, articles 2C and 2D; and annex VII, section I). The agreement on phasing out halons and CT by 2000 was a substantial accomplishment by any standard.

The parties also had to decide on a *base year* for calculating control levels for the newly added substances—CT, MC, and the new CFCs. They entered the meeting with two options: 1986 and 1989. The argument for the earlier base year was that it would result in larger interim reductions, and 1986 would be uniform with the first group of controlled chemicals.

However, 1989 was also consistent with the original protocol, in the sense that, like 1986 at the time, it was the year preceding imposition of controls. The parties decided that this was a good precedent, especially as future decisions on new chemicals would be increasingly remote from 1986 and the quality of more recent data would be better. The selection of 1989 also conveyed an incidental interim consolation to the producers and users of methyl chloroform, now unexpectedly facing a phaseout, as the parties explicitly acknowledged that there had been substantial growth in MC use between 1986 and 1989.

In sum, after the London revisions there were now five groups of controlled substances under the Montreal Protocol: the original five CFCs and three halons, plus ten new CFCs, carbon tetrachloride, and methyl chloroform. All of these were now scheduled for varying interim reductions and phaseout in 10 to 15 years, with clear indications that even these schedules could be accelerated on the basis of early reassessments. In addition, some 34 HCFCs were included in the protocol as "transitional substances," and the parties had signaled that HCFCs and "other halons" (which would be enumerated later by UNEP in a separate document) should be used only with discretion and would be subject to continuing scrutiny and future control.

The extent that the negotiators in London achieved a meaningful strengthening of controls is dramatically evident in Figure 13.1, which compares the effects of the original and the revised protocols on atmospheric chlorine concentrations. Not only is chlorine loading now held to a peak of slightly over 4 ppb, but the pre–ozone hole level of 2 ppb should be attained by 2075.

Fine-Tuning the Controls

The developing countries entered the debate on controls with fervor when it came to deciding on *procedures for future adjustments* in the reduction schedules. Here was a matter of equity. They had made clear early in working group discussions that they would no longer accept two-stage, semiweighted voting as introduced in the original protocol. The weighting provisions had in any case become moot, since article 5 countries now had the numbers to block decisions; for example, adjusting the schedules of already controlled substances under the original terms required ap-

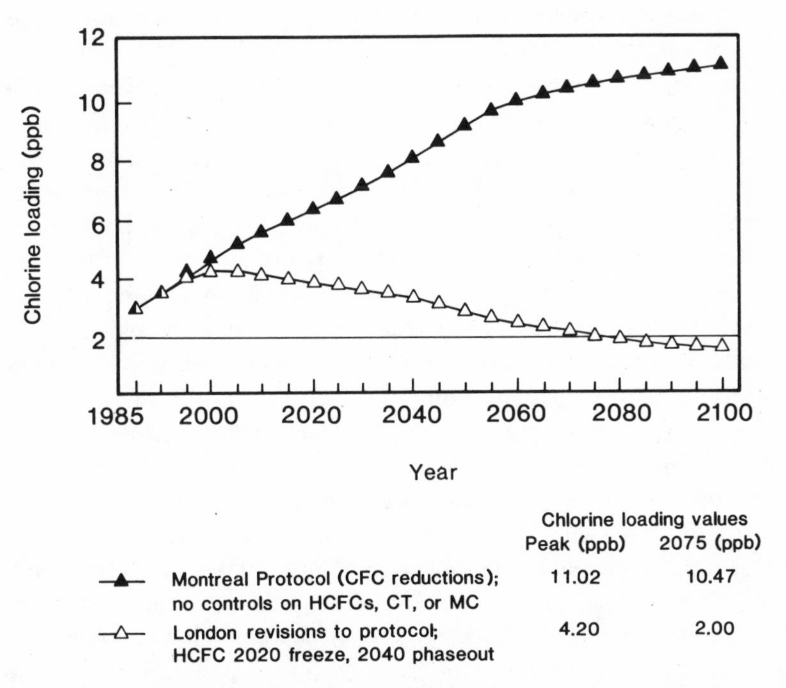

Figure 13.1 Effects of Montreal Protocol versus London revisions on atmospheric chlorine loading, 1985–2100. (Assumptions for both scenarios: intermediate reductions as specified in Montreal Protocol or London revisions; HCFC substitutes replace 30 percent of CFC reductions, have an atmospheric lifetime of 8 years and a chlorine-loading potential of 0.05; natural background chlorine loading of 0.6 ppb.) *Source:* Calculated by EPA.

proval by two-thirds of the parties representing at least 50 percent of combined total consumption of the substances (article 2, paragraph 9).

In any event, the developing nations insisted on deletion of the weighting condition. Further discussions in London evolved a principle of balance: changing the controls would now require a two-thirds majority of all parties, which would have to include submajorities of both the article 5 parties and the industrialized countries. This gave both North and South an effective veto. An initial proposal envisioned two-thirds submajorities also, but U.S. negotiators then realized that this arrangement would give the European Community, with its 12-vote bloc, undue weight in the industrialized-nation group. Hence, the final London text provided that future control revisions would require a two-thirds majority of all parties, representing separate simple majorities of both the article 5 parties and the others. Establishing simple majorities would also make future adjustments marginally easier to achieve.

Related to the control measures was a decision on the treatment of the *additional production allowance* at each reduction phase—the difference between consumption and permissible production. In the original protocol, this difference had been fixed at 10 percent for the first stages and 15 percent for the last, and had been specifically limited to purposes of satisfying "basic domestic needs" of article 5 parties and of "industrial rationalization" between parties. The London meeting dropped the latter as no longer necessary because of the change in article 2, paragraph 5, discussed below. Thus, the extra production leeway was restricted to the purpose of supplying developing countries during the phasedown period.

There was some interesting debate among developing countries on the desirable size of these allowances. Some, like Mexico, argued that a lower excess production limit for the industrialized countries (5–10 percent) would still be sufficient to supply CFCs to article 5 parties, because the new substitutes would be coming on line; incidentally, a lower percentage would also help protect Mexico's own CFC producers from import competition. But others, such as Malaysia, felt insecure about receiving adequate supplies of controlled substances during the phasedown period and hence argued for a higher percentage.

It was not entirely clear which alternative was better. If the extra production allowance proved inadequate, some article 5 parties might need to build their own capacity in substances that would soon be obsolete. On the other hand, if there was too much leeway for industrialized-country production, their producers might try to create new export markets for ozone-depleting chemicals. In the end, the parties decided to stay with the 10–15 percent formula for the present, and they commissioned the technology panel to analyze the probable requirements of developing nations for their basic domestic needs as well as the likely availability of supplies.

Also linked to the new controls were corresponding revisions in the *trade measures*. These provided for restrictions of trade with nonparties in the newly controlled substances analogous to those applying to the original CFCs and halons (annex II, article 4). In their quest for parity the developing nations unexpectedly strengthened the restrictions: the ban on exports of controlled substances to nonparties, which in the original protocol had applied only to exports from article 5 parties, was changed to apply to all parties. Another important addition to the protocol was that any existing party that did not accept the London amendment containing the newly controlled substances would be considered a nonparty with respect to those substances, and hence was subject to trade restrictions.

Finally, the European Community had proposed during earlier working group negotiations a change in the industrial rationalization clause to permit any party, not just small producers, to transfer production to another party, provided the secretariat was informed of the details and the combined total output of the two parties did not exceed the treaty limit (article 2, paragraph 5). This change was particularly important to the EC because of its emerging single market. It also made economic sense: without this provision, individual companies in a small or medium-sized country phasing down to uneconomical size could face a competitive disadvantage vis-à-vis producers in parties that had the capability of combining facilities within large internal markets.

Japan and the United States blocked approval of this clause until the final hours. Ultimately, agreement was linked to the resolution of other differences, as described above.

Elusive Data

It has been mentioned that Greenpeace, in particular, had drawn attention to the impossibility of independently verifying compliance with protocol provisions when some countries and companies continued to withhold complete and open data on their production, imports, and exports—

even though such specific reporting was required under the treaty. Support for data confidentiality was especially strong in the EC; at the First Meeting of Parties in Helsinki and earlier, the Community had attempted to weaken reporting requirements (see Chapter 10).

The problem of acquiring reliable data from traditionally secretive companies was acute. For example, the West German Federal Environment Agency observed in 1989 that "it is nearly impossible to obtain accurate data concerning the amounts of CFCs and other ozone-depleting substances that are nationally produced, exported and imported, in spite of the public attention." The German report further noted that, under EC regulations, "manufacturers are merely required to send their production figures to their national authorities and the EC, who will then publish them in summary form." The agency therefore cautioned that "national production and import figures provided by the industry in the Federal Republic of Germany and the European Community cannot be verified."[21]

West German environment authorities independently estimated German consumption of CFCs in 1986 (the critical base year for calculation of the Montreal Protocol controls) to have been within the rather broad range of 69,000–100,000 metric tons; by 1990, the figure had been fixed at 112,000 metric tons.[22] Knowledgeable officials believed that industry had deliberately understated 1986 data as long as they thought that indications of lower growth relative to 1985 might influence the negotiators toward weaker controls. This tactic backfired, however, when the treaty established 1986 as the base for future cutbacks, thereby placing a premium on a higher total; hence, the data had to be hastily "revised" upward. If data were unreliable in Germany, where public and government had been among the most forceful proponents of regulation, one can only wonder at the situation in some other countries.

UNEP had in fact become increasingly concerned over the attitude of parties toward reporting. In a report to the parties dated May 28, 1990, Executive Director Tolba again called attention to the inadequacy of data reporting by many governments. He noted that article 7 of the protocol, reiterated and clarified by a decision of the First Meeting of Parties, required each state party to provide annual data on production, exports, and imports for every individual controlled chemical, including trade data broken down between parties and nonparties. Such data were essential for implementing the protocol and for assessing compliance. Tolba emphasized that "it is crucial that the Parties report complete data."[23] Earlier

Tolba had gone so far as to suggest that, "if the situation with regard to incomplete data submission persists, it could possibly be considered noncompliance."[24]

The record was not good. UNEP had sent out several formal requests for data in 1989 and 1990. Of the 55 parties required to submit data by the end of May 1990, only 21—including the United States, Japan, and the Soviet Union—had provided complete reports; the European Community had also reported, but only as an aggregated unit. Sixteen parties had submitted incomplete data; 4 parties, all developing countries, stated that they had no data available; 4 reported that their information was included with that of another party. And 10 parties had not reported at all.[25]

Lack of adequate data from some developing nations was already beginning to affect their access to supplies of CFCs. Unless UNEP had sufficient information to establish that a country's consumption was less than 0.3 kilogram per capita, that country could not be certified as an article 5 party. Some industrialized-nation manufacturers were reportedly holding up exports because they could not, in these circumstances, receive the extra production rights allowed under article 2 for sales to article 5 parties.

In addition, there was a continuing problem with the European Community. The interests of the EC Commission on this issue coincided with those of some member states. The Commission preferred to submit aggregated data to symbolize and reinforce its federal role. And certain member governments, reflecting the wishes of their private industry, chose to submerge their data for purposes of safeguarding commercial confidentiality.

The EC Commission had only reluctantly agreed at Montreal and Helsinki to provide production, export, and import data separately by member country and by chemical. Only three EC members, however, reported complete data (Denmark, Luxembourg, and Spain), and only three submitted separate production figures (France, the Netherlands, and Spain). The Federal Republic of Germany and Ireland notified UNEP that their data were included in the EC submission. Four EC member states—Greece, Italy, Portugal, and the United Kingdom—had not reported any data at all to UNEP.

Other parties were concerned about this situation because it increased the difficulties of monitoring compliance of state parties, especially with the addition to the protocol of more stringent controls and new chemicals and phasedown steps. There was also the by-now-familiar matter of principle: each voting state party should comply with all protocol provisions, whether or not it was a member of a regional economic integration organization (see Chapter 7).

Many parties felt strongly that more open reporting was in the public interest and would ensure effective implementation of the protocol. The Nordic nations were partially successful in an attempt in London to lift the reporting secrecy that had been accepted in Helsinki and to open more data to public scrutiny. They persuaded the meeting of parties explicitly to confirm that consumption data (an aggregate, consisting of production, plus imports, minus exports) would not be treated as confidential, but they were unable to obtain the same treatment for separate production data. For its part, the EC tried to relax the principle further by proposing that collective reporting of production be allowed for members of a REIO; this proposal, however, was turned down by the working group.

The parties did in the end consent to a new clause permitting the EC, on behalf of its members, to submit aggregated data for exports and imports of controlled substances between the Community as a whole and nonmember states (annex II, article 7). The rationale for this provision was that after the advent of the European single market in 1992, official national trade data would not be available for individual EC member countries. European insiders, however, confided that national data could be maintained and that there was no legal reason—only a political one—for denying release of such data in fulfillment of a treaty obligation.

This provision enabled the EC to preserve the confidentiality of production figures for its individual members, since production could have been derived from the open consumption data plus national trade data. Thus, even after the London conference, data from individual EC member countries remained largely inaccessible. The Community had to provide open consumption data, but the protocol permitted EC member-country figures to be combined. In contrast, production data had to be reported by individual country, but these data were confidential and not subject to public scrutiny. Even delegations concerned about the data question, however, could not muster the energy to mount a major effort on this issue during the final urgency of forging consensus on the really big issues in London.

But the parties would have to return to the subject another day. Complete and reliable data are essential to verifying the claims of a company or nation that it is adhering to regulations.

In this connection, the parties at London also had to decide on specific procedures to deal with noncompliance. The parties had before them what Tolba characterized as an "encouragement-based approach" in the form of a proposal from the group of legal experts commissioned by the Helsinki meeting, as opposed to "more stringent and punitive" measures favored by some delegations.[26]

The legal group recommended establishment of an "implementation committee," consisting of five elected parties, to review complaints of noncompliance. The committee's only substantive function was to "consider" the available information and report to a meeting of parties, which could then determine what steps might be desirable to bring about full compliance. This recommendation found strongest support from the European Community and many developing countries.

Other parties, however, including the Nordic nations and the United States, felt that stronger medicine was needed. During the London discussions they tried to extend the trade restrictions to parties found in noncompliance, in order to preclude nations' ratifying the treaty merely to avoid the trade barriers. Although these efforts were unsuccessful, delegations favoring stronger noncompliance measures did prevent acceptance of the legal group's recommendation as the final word. Reflecting a general sense of dissatisfaction with the vagueness of the proposal, the London meeting adopted the procedure only on an interim basis. Moreover, it directed the legal group to return to the subject and to elaborate more specific terms of reference for the implementation committee as well as further procedures for treatment of parties in noncompliance. In a related decision, the parties also established a new ad hoc group to consider the problem of data reporting and to recommend possible solutions.

Innovative Funding

Notwithstanding the progress that had been made by the Open-Ended Working Group through its May session, details of the funding mechanism occupied a dominant portion of the London meetings. As the moment of decision approached for this precedent-setting step, most major donor countries wanted to ensure clear and predictable parameters for the new fund—the idea of which had met with strong resistance when it was initially proposed at the 1989 Helsinki meeting of parties.

At the outset, the U.S. statement on its revised policy, delivered by seasoned diplomat Richard J. Smith, both occasioned relief among the delegates and illuminated the remaining issues to be joined in London. By announcing its support for the principle of additionality, the United States overcame the biggest single obstacle to agreement. The United States made clear, however, that it would not buy into an open-ended commitment.

The United States sought to make explicit in the text of the protocol its view of the new fund as one of "limited and unique nature." Therefore, the U.S. negotiators proposed language specifying that (1) the ozone-depletion problem being addressed by the fund was scientifically established, (2) the funds would make a real difference in overcoming the problem, (3) the amounts needed were predictable, and (4) the financial mechanism "is without prejudice to any future arrangements that may be developed with respect to other environmental issues."[27]

Reflecting the political importance of the last point for the United States, the U.S. delegation insisted on its inclusion in the body of the new article establishing the financial mechanism (annex II, article 10). In accepting this proposal, many delegates noted privately that "without prejudice" was more neutral and less restrictive than "without precedent" would have been. The other three points were placed in a preambular clause to the amendment; although they thus had no practical effect on the treaty, their inclusion here satisfied U.S. domestic concerns. Even so, Norway and other delegations used their acceptance of the clause for leverage in the final hours by linking their approval to U.S. movement toward stricter controls over methyl chloroform.

Substantively, the U.S. position on the fund specified that it should be operated and administered by the World Bank, with oversight vested in an executive committee to be established by the parties. As the largest prospective donor to the fund, the United States announced that it expected a permanent seat on the committee. Furthermore, it proposed that committee decisions be reached through a voting mechanism that was weighted to reflect size of contributions.

Although the frank request for a permanent seat raised some hackles, most delegations acknowledged its fairness, and a way was found to satisfy the United States without formally defining transcendence (other governments, it seemed, also had principles about setting precedents). An executive committee was formed with balanced representation of seven members each from article 5 parties and industrialized countries. It was left to each group to agree informally on how its seats would be allocated. The donor countries decided that representation on the committee would rotate within separate "regional" groups; for example, the three Nordic states would alternate in occupying one seat, and EC members would share two. With seven "regions" established, it was clear that the United States constituted a region unto itself.

The executive committee was charged with developing policies and was given extensive supervisory, administrative, and operational authority (annex IV, appendix II). This included preparing the fund's budget for approval by the parties, allocating funds among the cooperating agencies (the World Bank, the United Nations Development Programme, UNEP, and possibly others), evaluating programs, and approving projects of over $500,000. The initial members of the committee were Canada, the Federal Republic of Germany, Finland, the Netherlands, Japan, the Soviet Union, and the United States, representing the industrialized nations; and Brazil, Egypt, Ghana, Jordan, Malaysia, Mexico, and Venezuela, representing the article 5 parties (China and India were not parties, and so were not yet eligible). Finland and Mexico were elected as chair and vice-chair, respectively, for the first year.

There was some North-South maneuvering for control over decision making in the new fund, as determined by its voting procedures. Some major donors, drawing on article 13, paragraph 2, of the original treaty ("The Parties, at their first meeting, shall adopt by consensus financial rules for the operation of this Protocol"), proposed that decisions dealing with the fund be subject to consensus. This was a brave try, but the article, which had specified procedures only for the first meeting of parties, could not be stretched in interpretation to apply to the new fund at the second meeting.

For their part, developing-country legal experts referred to the previously agreed rules of procedure on general decisions, as well as to the protocol amendment procedure (Vienna Convention, article 9), both of which provided for a straightforward two-thirds majority rule. Applying those rules would, at some point in the future, give them control of the new fund.

The resultant compromise drew on the new, balanced solution that had just been agreed upon for the voting on controls: it was decided that a two-thirds majority, comprising separate simple majorities among North and South, would apply both to votes of the 14-member executive committee and to votes of the parties as a whole concerning the financial mechanism. Thus, both donors and recipients had potential blocking power.

The seeming intricacy of these procedures must be weighed against the obvious desire of all participants, as reflected throughout the negotiations, to achieve consensus without voting; no vote, in fact, was ever taken at Montreal, London, or any other meeting. But North-South balance was clearly the order of the day.

Because the amendment establishing the new fund could not enter into force until 1992, the parties adopted an "interim financial mechanism" to begin functioning on January 1, 1991. The negotiators of the donor governments were successful in avoiding the location of the new mechanism with UNEP in Nairobi and the designation of UNEP as administrator of the fund. Such an arrangement would have entailed a standard UN 13 percent "administrative support charge." Instead, the Canadian government came forward with an irresistible offer to host the executive committee during the interim period and to assume the related administrative costs. The matter was not definitively settled at London, but at the initial meeting of the executive committee in September 1990 in Montreal, the Canadian proposal was formally accepted and Montreal was, appropriately, designated as the seat of the new fund and secretariat. In a decision that was simultaneously consolatory and economical, the committee augmented UNEP's symbolic role by accepting Tolba's offer for UNEP to perform a banking function—basically issuing and receiving checks—at no cost.

The parties in London also ultimately delegated a paramount role within the fund to the World Bank, which in earlier draft proposals had appeared as little more than a cashier's window. In the terms of reference approved in London, UNEP would pursue "political promotion of the objectives of the Protocol," as well as research, data collection, and clearinghouse functions. UNDP would take charge of the feasibility studies and other technical assistance activities. Other multilateral agencies, including regional development banks, could be invited by the executive committee to cooperate with the fund. But the World Bank, and specifically the president of the bank, was clearly designated as the administrator and manager of the central function of the fund: financing projects and programs to meet the incremental costs of article 5 parties (annex IV, appendix IV).

Results from initial country studies were presented in London and proved helpful in meeting the U.S. criterion of establishing some predictability concerning the magnitude of needed resources. The report on China commissioned in 1987 by the Open-Ended Working Group was of particular importance because of the size of China's market. Careful analysis by an international team working in China with experts there revealed that, contrary to initial estimates of up to $1 billion, approximately $42 million in incremental costs would be needed over the next three years to help China toward phaseout of controlled substances.[28] These estimates suggested that replacement of aerosols, recovery equipment, demonstration plants, and, especially, new technology could substantially reduce prospective usage of CFCs and halons. The Chinese government accepted the report and also indicated that it did not plan to make use of the 10-year grace period allowed by the treaty to increase consumption of controlled substances; rather, it strongly preferred to move quickly to new technologies and avoid contributing further to ozone layer damage.

Similar studies presented in London indicated that a total phaseout over 20 years would entail $30–38 million for Egypt and $64–119 million for Mexico. On the basis of discussion of these and other studies still under way, the parties determined that, for the initial three years of the new fund, $160 million would be a reasonable budget, plus an additional $80 million were China and India soon to become parties; this sum included approximately $20 million for the operation of the clearinghouse function.

With the amount of the interim fund decided, the issue of voluntary versus mandatory contributions by industrialized countries was solved by adroit drafting. Extremely careful reading of the new article 10 text would reveal that contributions were not obligatory, but neither did the word *voluntary* appear. The fact that a detailed "scale of contributions," expressed in percentages to two decimals, was accepted and appended, did convey the impression of at least a tacit commitment.

The formula for burden-sharing did occasion some discussion among the donor governments. The United States preferred the UN assessment scale, which automatically capped its contribution at 25 percent, rather than a formula based on 1986 consumption, which for the United States would have been around 30 percent. EC Commission representatives, however, complained that the UN scale, which was based on individual countries, yielded a total of nearly 35 percent for the Community as a whole, and thus would establish an unfair relationship to the United States. But the EC's reluctance to provide individual-nation consumption data proved costly, because the Commission could offer no alternative. There was overwhelming endorsement of the UN scale, which had the advantage of considerable precedence. Prime Minister Thatcher and U.S. EPA Administrator William Reilly promptly announced their governments' willingness to participate on this basis, which in the case of the United States amounted to $40–60 million for the three-year period.

It was also agreed in London that a donor country could count bilateral aid of up to 20 percent against its putative contribution to the multilateral fund, provided that this aid was additional and specifically in accord with the fund's criteria. However, a recipient country had the right to request that all its needs be met solely from the resources of the multilateral fund.

In sum, the agreed charter for an ozone fund was an exceptional innovation in the realm of multilateral cooperation. Procedures and terms of reference had been devised incorporating delicate checks and balances among donors and recipients. The parties to the protocol exercised ultimate authority over the new financial mechanism. The 14-member executive committee represented the parties in overall supervision and administration of the fund, which would have only a small central secretariat. The three principal collaborating multilateral institutions would be assigned specific responsibilities under interagency agreements with the executive committee. Regular consultations among the cooperating agencies were planned, and the heads of the organizations would be invited to meet at least annually with the executive committee. An experiment had been launched with important implications for future approaches to global problems requiring North-South cooperation.

North-South Endgame

Difficult as the deliberations over details of the financial mechanism had been, they were less laborious than the parallel negotiations over access to technology and the obligations of developing countries under the protocol. This issue occasioned several late-night sessions of the London working group, and even the ministers subsequently could not resolve it until the fate of the entire agreement hung in the balance.

The problem was that both sides wanted ironclad guarantees that were understandable from their perspective, but that were also mutually exclusive. Developing countries wanted to build into the protocol an assurance that, if they did not receive sufficient financial and technical help, they would not be obligated to implement the reduction schedules. For their part, industrialized nations recognized the reality of a linkage between external aid and the capability to renounce usage of CFCs and the other substances. But the donor countries would agree neither to a blank check nor to a situation in which any party could unilaterally claim that it had received insufficient aid and therefore would not fulfill treaty obligations. Underlying the debate was a difference in degree of confidence in the ability of the international market mechanism to transfer modern technology fairly and effectively.

The developing-country case at this stage was expressed most forcefully by India's environment minister, Maneka Gandhi, daughter-in-law of the late prime minister Indira Gandhi. Major donor-government spokesmen included EPA's Reilly and U.K. environment secretary Chris Patten. Mexico's sedulous Ambassador Mateos proved to be an indispensable mediator in the debate.

Gandhi advanced the demand for mandatory technology transfer in the starkest terms. She stated that money was irrelevant without accompanying access to knowledge: "The whole 21st century's survival will be based on . . . knowledge." She was skeptical of claims by Western governments that they were unable to guarantee transfer of technology because the intellectual property rights were in private hands; she noted that governments intervene in the operations of the market all the time: "Either you [sell us] the technology or you change your laws or you change your patent rights . . . Start working on it!"[29]

The delegation heads of Brazil, China, Malaysia, and others joined Gandhi in expressing apprehension that Western enterprises could use this situation to derive "exorbitant" profits.[30] Malaysia's Minister of Science, Technology, and Environment D. A. S. Yong declared that denying access to modern technology amounted to "environmental colonialism." Gandhi added that "we have a problem [about] turning into a client state."[31]

The protocol language demanded by the developing-country bloc would have, in effect, qualified their obligations by permitting unilateral abrogation of treaty requirements if an article 5 party decided it had not received enough help: "the obligation . . . to comply . . . will be subject to" adequate financial assistance and "preferential and non-commercial" transfer of technology.[32]

The debate took on a strident tone as stalemate continued. Gandhi went to the press, stating that neither India nor China would ratify the treaty and threatening to assist article 5 parties in withdrawing from the protocol. India was, however, beginning to risk isolation. Privately, representatives of industrialized countries spoke of "environmental blackmail." Several developing-country delegations, including Ghana and

text continues on page 196

Table 13.1 Summary of 1987 Montreal Protocol and 1990 revisions

	1987 Montreal Protocol	
Topic	Provision	Location
Chemicals		
Chlorofluorocarbons (CFCs) 11, 12, 113, 114, 115	Freeze at 1986 levels beginning in mid-1989	Art. 2, para. 1
	20% reduction beginning in mid-1993	Art. 2, para. 3
	50% reduction beginning in mid-1998	Art. 2, para. 4
Halons 1211, 1301, 2402	Freeze at 1986 levels in 1992	Art. 2, para. 2
10 other fully halogenated CFCs	Not covered	
Carbon tetrachloride	Not covered	
Methyl chloroform	Not covered	
Other halons	Not covered	
Hydrochlorofluorocarbons (HCFCs)	Not covered	

1990 London revisions	
Provision	Location
No change 50% reduction in 1995 85% reduction in 1997 Phaseout in 2000 Schedule to be reassessed in 1992 with objective of accelerating reductions.	Art. 2A
No change 50% reduction in 1995 Phaseout in 2000 Parties will determine in 1992, with subsequent review, whether any essential uses should be exempt from reductions.	Art. 2B
20% reduction from 1989 levels in 1993 85% reduction in 1997 Phaseout in 2000	Art. 2C
85% reduction from 1989 levels in 1995 Phaseout in 2000	Art. 2D
Freeze at 1989 levels in 1993 30% reduction in 1995 70% reduction in 2000 Phaseout in 2005 Schedule to be reassessed in 1992 with objective of accelerating reductions.	Art. 2E
Nonbinding resolution discourages usage, requests reporting on production and consumption.	Annex VII
Mandatory reporting on production and consumption.	Art. 7
Nonbinding resolution calls for phaseout no later than 2040, and if possible by 2020, with regular reassessments.	Annex VII

Table 13.1 (*continued*)

Topic	1987 Montreal Protocol Provision	Location
Developing countries' obligations	Increase in consumption of controlled substances allowed up to 0.3 kilogram per capita for 10 years in order to meet "basic domestic needs." After 10 years, the reduction schedule must be followed.	Art. 5, para. 1
Financial assistance	Parties will "facilitate" bilateral and multilateral aid to developing countries.	Art. 5, para. 3
Technology transfer	Parties will "facilitate access" to technology by developing countries.	Art. 5, para. 2
	Parties will promote exchange of information and technical assistance.	Arts. 9 and 10
Review of control measures	Controls will be evaluated on basis of scientific, environmental, economic, and technological assessments beginning in 1989 and at least every four years thereafter.	Art. 6
Voting		
Canceling 50% reduction of CFCs	Requires approval by two-thirds of parties representing at least two-thirds of consumption of all parties.	Art. 2, para. 4
Adjustments in reduction of already controlled substances	Requires approval by two-thirds of parties representing at least 50% of consumption of all parties. Binding on all parties.	Art. 2, para. 9

1990 London revisions	
Provision	Location
No change for original CFCs and halons. For new controlled substances, limit is 0.2 kilogram per capita.	Art. 5, para. 2
Article 5 party may appeal to meeting of parties if financial aid and technology transfer (see below) are inadequate to enable it to comply with treaty obligations.	Art. 5, paras. 5–9
Multilateral fund, administered by World Bank under policy control of parties, will finance incremental costs to enable compliance with controls. Feasibility studies and technical assistance will also be financed. Initial 3-year budget set at $160–240 million.	Art. 10; annex IV; app. IV
Parties will "take every practicable step" to transfer technology to article 5 parties "under fair and most favourable conditions."	Art. 10A
No change	
Not applicable	
Requires approval by two-thirds of parties comprising separate majorities of developing countries (article 5 parties) and industrialized countries. Binding on all parties.	Art. 2, para. 9

Table 13.1 *(continued)*

Topic	1987 Montreal Protocol	
	Provision	Location
Addition of new substances	Amendment to protocol; requires approval by two-thirds of parties. Binding only for parties that ratify amendment.	Art. 2, para. 10
Decisions on financial mechanism	Not applicable	
Trade restrictions		
Exports to nonparties of controlled substances in bulk	Prohibited for exports from developing countries, beginning in 1993.	Art. 4, para. 2
Imports from nonparties of controlled substances in bulk	Prohibited beginning in 1990.	Art. 4, para. 1
Imports from nonparties of products containing controlled substances	Prohibited beginning in 1993.	Art. 4, para. 3
Imports from nonparties of products made with controlled substances	By 1994, parties will determine feasibility of ban.	Art. 4, para. 4
Noncompliance procedures	Procedures for determining noncompliance and for treatment of parties in noncompliance to be decided.	Art. 8
Entry into force	January 1, 1989, provided ratifications are received from at least 11 nations, representing at least two-thirds of estimated 1986 global consumption of controlled substances.	Art. 16

Sources: For 1987 Montreal Protocol, Appendix B; for 1990 London revisions, Appendix C.

1990 London revisions	
Provision	Location
No change	
Requires approval by two-thirds of parties comprising separate majorities of developing countries (article 5 parties) and industrialized countries.	Art. 10, para. 9
Prohibited for exports from all parties. For new controlled substances, prohibited beginning in 1993.	Art. 4, paras. 2 and 2 *bis*
For new controlled substances, prohibited beginning in 1993.	Art. 4, para. 1 *bis*
For new controlled substances, prohibited beginning in 1996.	Art. 4, para. 3 *bis*
For new controlled substances, by 1997 parties will determine feasibility of ban.	Art. 4, para. 4 *bis*
Implementation committee established to review complaints.	Annex III
Parties ask legal experts to develop more detailed procedures.	(decision)
January 1, 1992, provided that at least 20 parties ratify the amendment.	Annex II, art. 2

some Latin American nations, expressed belief that market forces, reinforced by the emerging language in the protocol, would provide sufficient assurance of access to fairly priced technology. The Chinese delegation head, Wang Yangzu, in a gesture widely interpreted as a signal to Gandhi that China was capable of speaking for itself, called his own press conference late on June 28 to declare satisfaction with the progress of the negotiation and to indicate that China was now ready to accede to the Montreal Protocol.[33]

The resolution of the impasse was ingenious, realistic, and acceptable to all; on June 29, India joined China in announcing that its delegation would now recommend adherence to the revised protocol.

Industrialized countries affirmed their commitment to "take every practicable step . . . to ensure that the best available, environmentally safe substitutes and related technologies are expeditiously transferred [to article 5 parties] . . . under fair and most favourable conditions" (annex II, article 10A).

The protocol text now formally acknowledged that "developing the capacity [of article 5 parties] to fulfil the obligations . . . will depend upon the effective implementation" of the financial mechanism and technology transfer (annex II, article 5). This statement represented the reality, but it did not go so far as to release parties from their treaty obligations.

A fair solution was also devised for the knotty problem of who was to decide that assistance had been insufficient. The amended article 5 provided that, if a developing country believed itself unable to comply with control measures, it could notify the secretariat, and the parties would consider appropriate action at their next formal meeting; in the interim, there would be no noncompliance procedures invoked against the notifying party. Decisions by the parties on these matters would be governed by the same balanced North-South voting procedure that applied to the fund and to adjustments of controls.

A Dynamic Precedent

The Montreal Protocol, as its designers had intended, was responding dynamically to changed conditions—and was setting valuable precedents along the way. A fatigued but visibly relieved Mostafa Tolba proclaimed on the night of June 29 that the Second Meeting of Parties had not merely strengthened a treaty, but had written "a new chapter in the history of international relations."[34]

Even the atmosphere of the conference had been unique. It was a "meeting of parties," but two of the most important actors—China and India—had been nonparties. It is noteworthy that the negotiations succeeded in adopting all decisions by consensus—including the 42 delegations from governments not yet party to the protocol; the elaborate voting procedures were never invoked.

Exemplifying the conciliatory spirit was a vignette from the last night of the London working group, when it was desperately striving to complete its work before the arrival of ministers the following morning. All translators had to leave at 10:00 P.M. because of union contract obligations. Mexico's Mateos, in the chair, having previously conducted the plenary negotiations in his native Spanish, contravened all United Nations rules by appealing to delegations to continue their work in the English language. Several delegations reluctantly agreed, but a *frisson* of apprehension swept through the hall as France, which by tradition scrupulously holds to linguistic parity, claimed the right to speak. The French representative declared that, although the world acknowledged the elegance, precision, and importance of French as the language of diplomacy, in the current situation his delegation believed that protecting the ozone layer was even more important: therefore, the French delegation concurred in the use of English for this session. An unusual wave of applause greeted this gracious gesture, and the negotiation continued.

In a reflection of political developments outside the conference hall, cooperation among Western, Soviet, and Eastern European delegations was exceptional throughout the deliberations. East and West German environment ministers used the occasion for informal consultations on the forthcoming merger of their ministries under German reunification, and in many respects the two delegations already functioned as one.

Another unusual feature was the formal participation in the meeting of parties by a delegation of Australian youth, symbolizing the fact that the next generation would experience the consequences of ozone layer depletion to a much greater degree than any of the negotiators. Addressing the assembled ministers, a 17-year-old girl received a standing ovation when she pleaded for a strong treaty and declared that "our fate lies within your square brackets."

Finally, recalling the controversies over ozone theories going back to the 1970s (see Chapter 3), there was a certain delicious irony, together with a sense of historical symmetry, in a decision by the parties to instruct the science assessment panel to reexamine the effects on the ozone layer of emissions from high-altitude aircraft, heavy rockets, and space shuttles.

* * *

The negotiators in London were acutely aware of the precedents they were setting for approaches to other global environmental issues. In many ways, the debates over the ozone protocol were serving as a preview to the now-scheduled negotiations for a climate convention and the 1992 United Nations Conference on Environment and Development. Many ministers at the meeting, including Prime Minister Thatcher, explicitly stated that a successful outcome to the protocol negotiations would pave the way for international agreement on measures to address global warming.[35] This view was echoed by the spokesman for a prominent U.S. industry association.[36]

The parties and nonparties to the Montreal Protocol had accomplished far more than significantly strengthening controls over ozone-depleting substances: they had created the first financial mechanism dedicated to protection of the global environment, and, for the first time, the governments of industrialized nations had accepted a responsibility to help developing countries with modern technology.

The president of the London conference, U.K. environment secretary Patten, concluded that the Montreal Protocol and its process would become "the model for . . . future environmental diplomacy."[37]

Looking Ahead: A New Global Diplomacy

14

Perhaps the most poignant image of our time is that of Earth as seen by the space voyagers: a blue sphere, shimmering with life and light, alone and unique in the cold vastness of the cosmos. From this perspective, the maps of geopolitics vanish, and the underlying interconnectedness of all the components of this extraordinary living system—animal, plant, water, land, and atmosphere—becomes strikingly evident.

Experimenting with Planet Earth

Humanity has learned that the activities of modern industrial economies, driven by consumer demands and burgeoning populations, can alter delicate natural balances. We can no longer pretend that nothing will happen as the planet is subjected to billions of tons of pollutants. It is not that Earth itself is necessarily fragile. It may be, rather, that our own tenure turns out to be less secure and less inevitable as a consequence of the planet's responses to unnatural conditions created by human actions.

The Antarctic ozone hole conveyed a warning. Nature is capable of producing unpleasant surprises: even seemingly small interferences—in this case, an increase in stratospheric chlorine concentrations of a little more than one part per billion—can trigger dramatic and sudden reactions. Recent experience with the forests of Europe and North America indicates that other areas may also have unforeseen thresholds beyond which natural processes are unable to absorb the assaults of contemporary society. The world may not have the luxury of early warning signals before an irreversible collapse occurs in some other part of the planet's ecosystem.

A 1986 report by the National Aeronautics and Space Administration observed that "we are conducting one giant experiment on a global scale by increasing the concentrations of trace gases in the atmosphere without knowing the environmental consequences."[1] This statement echoed almost verbatim the warning three decades earlier of Roger Revelle and Hans Suess, two scientists at the Scripps Institution of Oceanography. Concerned about the implications for global warming of the rapid accumulation of carbon dioxide in the atmosphere resulting from fossil fuel combustion, they concluded in 1957 that "human beings are now carrying out a large scale geophysical experiment of a kind that could not have happened in the past."[2]

Worries over climate change have moved in recent years from far-sighted members of the scientific community to the mainstream of international politics. A consensus has developed that increasing atmospheric concentrations of carbon dioxide, along with other gases such as chlorofluorocarbons, methane, and nitrous oxide, could within a few decades cause temperatures to rise more rapidly, and to higher levels, than occurred in previous natural cycles going back for hundreds of thousands of years.[3] Such a development could have portentous implications, including widespread coastal inundation as a result of rising sea levels, major declines in agricultural productivity linked to changed rainfall patterns and increased soil erosion, greater incidence and severity of hurricanes and other storms, extinction of many plant and animal species, and creation of tens of millions of environmental refugees. There could be profound economic and social dislocations throughout the world.

Excessive greenhouse warming is an even more complex issue than depletion of the ozone layer. It involves more contributing factors, more wide-ranging and uncertain consequences, and more economically painful choices. Modern society is faced with the prospect of rethinking many ways of doing business to which it has become accustomed. Efforts to limit the magnitude and rate of temperature rise, and to adapt to its effects, will undoubtedly imply changes in energy, industry, agriculture, development, and population policies, as well as in consumption habits.

The new environmental threats to national and planetary security—of which climate change appears to be the most far-reaching—challenge both traditional science and diplomacy. A new science has evolved in recent years, made possible by advances in computer modeling, satellites, and measurement technologies. Known as earth systems science, this discipline attempts to integrate chemistry, physics, biology, geology, anthropology, meteorology, oceanography, and other subjects in order to understand more fully the interrelated forces that govern this planet. In 1986, the International Council of Scientific Unions (ICSU) launched a long-term interdisciplinary initiative that has been described as the biggest international scientific effort ever organized. Its purpose is to develop new insights into planetary processes and the ways they are being affected by human activities. Designated as the International Geosphere-Biosphere Programme, it was adopted unanimously by the more than 70 national academies and 20 international scientific unions that form the membership of ICSU—a manifestation of universal concern about the seriousness of the problems.[4]

Dilemmas for Policy: Striking a Balance

Because cooperation among sovereign states is essential for developing effective policies to address these issues, the new science requires an analogue in the realm of international relations.

The negotiators of the Vienna Convention and the Montreal Protocol faced issues similar to those raised by potential climate change. The science was uncertain, and the predicted harmful effects, though grave, were remote and unproved. Entrenched industrial interests claimed that new government regulations would cause immense economic and social dislocations. Technological solutions were either nonexistent or were considered unacceptable by most major participating governments.

Under these conditions, some governments allowed commercial self-interest to influence their scientific positions and used the scientific uncertainty as an excuse for delaying difficult decisions. Many political leaders were long prepared to accept future environmental risks rather than to impose the certain short-term costs entailed in limiting use of products seen as necessary to modern standards of living. Short-range political and economic concerns were thus formidable obstacles to cooperative international action based on the ozone-depletion theory.

Government policymakers face a dilemma in attempting to deal with these new environmental challenges. Premature actions or regulations based on imprecise and possibly incorrect theories and data can incur costs that later turn out to be unnecessary. But postponing a decision also might not be cost free. Waiting for more complete evidence can run the risk of acting too late to prevent major and possibly irreversible damage. In this eventuality, the future economic and social costs could be much higher, perhaps even catastrophic.

Unfortunately, the current state of economics is not helpful in analyzing such a situation. Its traditional methods of measuring income and growth appear increasingly irrelevant in the modern world: the more chlorofluorocarbons a country produces, for example, the greater is the growth in its gross national product. And the economist's prescription for coping with future monetary flows is inadequate for responding to large but distant dangers: present-value discounting can reduce even huge long-term costs to insignificance. Thus, the application by policymakers and investors of the tools of conventional economics may result in ecologically very misleading decisions.[5]

The history of the efforts to protect the ozone layer clearly demonstrates the crucial role played by industry in developing and implementing international environmental policy. Although U.S. initiatives were instrumental in first reducing CFC emissions in the 1970s, in achieving the Montreal Protocol in 1987, and in strengthening it in 1990, on four occasions during the ozone negotiations the U.S. government either reversed or came close to reversing its position.[6] It is noteworthy that on the three occasions since 1985, pragmatically oriented American industry forces intervened *in favor of* the international regulatory regime.

The response of industry to an environmental problem is conditioned by a complex of considerations and pressures generated by the market, which is itself directly influenced by consumer preferences and by government regulatory actions; and underlying these immediate influences may be educational activities of the media and environmental organizations. The personal values of corporate leadership and stockholders, as well as general societal attitudes toward a given environmental issue, will also affect industry reactions.

In the early years of the ozone history, both American and European industrialists were resolutely opposed to controls over CFCs. It was almost as if business leaders simply could not bring themselves to believe that these seemingly ideal chemicals, with so many benefits to society, could be capable of inflicting a remote outrage on the environment. American industry, however, through the Chemical Manufacturers Association, was consistently committed to resolving the uncertainties raised by the scientists, even if the results were to prove unpleasant; industry strongly promoted international scientific research and the 1985 Vienna Convention. The position of American industry toward international regulation was also conditioned by the desire for a "level playing field" with their European competitors.

During the negotiations before both the Montreal conference in 1987 and the London meeting of parties in 1990, industrialists almost inevitably complained that the proposed controls were too harsh and too costly to implement, while environmental groups argued that they did not go far enough. The treaty negotiators had to seek a balance between being too soft on industry and thereby running greater risks to the ozone layer, and demanding too much. Economically or technically unrealistic controls could generate severe dislocations. Bitter industry resistance, loss of jobs, long court battles to delay implementation, consumer backlash, noncompliance, could all delay the needed structural changes and substitute products.

Industry needs a firm signal and a stable regulatory environment to give it the confidence to invest in new and safer technologies; this consideration emerged very clearly in the debate during 1989 and 1990 over the HCFC substitute chemicals. Sometimes, however, it may be necessary to overcome severe inertia or preoccupation with near-term profits. Control measures may, therefore, need to be more stringent than industry would like, in order to provide a spur to environmentally responsible investment. As West German environment minister Klaus Töpfer (a member of the conservative Christian Democratic party) put it in June 1990: "I am absolutely convinced that if you give a clear cut timetable, it will stimulate industry to come up with substitutes. But if more time is allowed, I really believe they will take more time."[7]

In these circumstances, informed consumers can have a decisive influence. In the case of CFCs, aroused American consumers, through their changed purchasing habits, brought down the U.S. aerosol spray market by two-thirds in the late 1970s, and thereby stimulated competitive development of substitutes even before government regulations were in place. In 1988, threatened consumer boycotts against CFC aerosols in the United Kingdom, promoted by environmental groups, convinced British producers to phase out these products. Citizens of Taiwan blocked the erection of a CFC production facility in 1989. Public opinion in the Federal Republic of Germany had reached such a state by 1990 that it created the conditions for the earliest phaseout of ozone-depleting substances by any major producing country—with the voluntary cooperation of private industry.

Even though large companies have substantial resources available for public relations and advertising, it is no longer worthwhile for them to try to resist when consumer feelings become sufficiently intense. And this same public interest can help to discourage possible temptations to evade environmental regulations, whether by exploiting loopholes or by moving to other countries.

Unfortunately, amid the cacophony of competing claims and data, there are no ready formulas for government policymakers to apply in achieving the right balance of risk insurance. In the ozone negotiations, tentative proposals were advanced, analyzed, and debated with both industrialists and environmentalists. In the end, the negotiators had to rely on some combination of experience and instinct when it came to drawing the line. That neither industry nor the environmental groups were ever entirely satisfied with the results is perhaps inevitable and may even be an indicator of a reasonable outcome.

Lessons for a New Diplomacy

The international community was successful in its approach to the problem of protecting the stratospheric ozone layer. The experience gained from this process suggests several elements of a new kind of diplomacy for addressing such similar global ecological threats as greenhouse warming.

Scientists must play an unaccustomed but critical role in international environmental negotiations. Without modern science and technology, the world would have remained unaware of the danger to ozone until it was too late. Science became the driving force behind ozone policy. The formation of a commonly accepted body of data and analyses and the narrowing of ranges of uncertainty were prerequisites to a political solution among negotiating parties initially far apart. In effect, a community of scientists from many nations, committed to scientific objectivity, developed through their research an interest in protecting the planet's ozone layer that transcended divergent national interests. In this process, the scientists had to assume some responsibility for relating the implications of their findings to alternative remedial strategies. Close collaboration between scientists and key government officials who also became convinced of the long-term dangers ultimately prevailed over the more parochial and short-run interests of some national politicians and industrialists.

Governments may have to act while there is still scientific uncertainty, responsibly balancing the risks and costs of acting or not acting. By the time the evidence on such issues as the ozone layer and climate change is beyond all dispute, the damage may be irreversible, and it may be too late to forestall serious harm to human life and draconian costs to society. Politicians must therefore resist a tendency to lend too much credence to self-serving economic interests that demand scientific certainty, maintain that dangers are remote and unlikely, and insist that the costs of changing their ways are astronomical. The signatories at Montreal knowingly imposed substantial short-run economic dislocations even though the evidence was incomplete; the prudence of their decision was demonstrated when the scientific models turned out to have underestimated the effects of CFCs on ozone. Governments must sponsor the needed research and act responsibly on the basis of often equivocal results. Unfortunately, the current tools of economic analysis are inadequate aids in this task and can even be deceptive indicators; they are in urgent need of reform.

Educating and mobilizing public opinion are essential to generate pressure on hesitant governments and private companies. The interest of the media in the ozone issue and the collaboration with television and press by diplomats, environmental groups, and legislators had a major influence on governmental decisions and on the international negotiations. Concerned consumers brought about the collapse of the CFC aerosol market. And in their educational efforts, the proponents of ozone layer protection generally avoided invoking apocalypse and resisted temptations to overstate their case in order to capture public attention; exaggerated pronouncements could have damaged credibility and provided ammunition to those interest groups that wanted to delay action.

Multilateral diplomacy, involving coordinated negotiations among many governments, is essential when the issues have planetary consequences. The manifold activities of an international organization—the United Nations Environment Programme—were crucial in promoting a global approach to the protection of stratospheric ozone. UNEP coordinated research, informed governments and world public opinion, and played an indispensable catalytic and mediating role during the negotiation and implementation of the protocol.

Strong leadership by a major country can be a significant force for developing international consensus. The U.S. government reflected its concerns over the fate of the ozone layer through stimulating and supporting both American and international scientific research. Then, convinced of the dangers, it undertook extensive diplomatic and scientific initiatives to promote an ozone protection plan to other countries, many of which were initially hostile or indifferent to the idea. As the largest emitter of both ozone-destroying chemicals and greenhouse gases, the United States has enormous potential to influence the policy considerations of other governments in favor of environmental protection. In fact, because of the geographic size and population of the United States, its economic and scientific strength, and its international interests and influence, progress in addressing global environmental problems can probably not be achieved without American leadership.

It may be desirable for a leading country or group of countries to take preemptive environmental protection measures in advance of a global agreement. When influential governments make such a commitment, they legitimate change and thereby undercut the arguments of those who insist that change is impossible. Preemptive actions can also support moral suasion in encouraging future participation by other countries. In addition, action by major countries can slow dangerous trends and hence buy time for future negotiations and for development of technological solutions. The 1978 U.S. ban on aerosols relieved pressure on the ozone layer and lent greater authority to the U.S. government when it subsequently campaigned for even more stringent worldwide measures. Although environmental controls might conceivably affect a country's international competitiveness in the short run, they might also, by stimulating research into alternative technologies, give that country's industry a head start on the future.

The private sector—including citizens' groups and industry—is very much involved in the new diplomacy. The activities of both environmental organizations and private industry in undertaking research, lobbying governments, and influencing public opinion definitely affected the international debate on the ozone issue. A major by-product of the ozone negotiations was the development of closer relations among hitherto separate environmental groups around the world, reflected in their new and unaccustomed cooperation at conferences on the ozone layer and climate change in 1989 and 1990. Environmental organizations can also play an informal future watchdog role in monitoring compliance by industry with internationally agreed-upon commitments. For their part, industrialists are becoming aware that their corporate image is increasingly affected by environmental issues. The intellectual and financial resources of the corporate sector are, moreover, essential for developing the required technological solutions. These groups will expect to be involved in establishing policies and negotiating international accords on climate and other emerging global ecological issues.

Economic and structural inequalities among countries must be adequately reflected in any international regulatory regime. In the longer run, the developing countries, with their huge and growing populations, could undermine efforts to protect the global environment. As a consequence of the ozone issue, the richer nations for the first time acknowledged a responsibility to help developing countries to implement needed environmental policies without sacrificing aspirations for improved standards of living. The Montreal Protocol broke new ground with the creation of the multilateral ozone fund and the commitments for technology transfer, while illuminating the issues that must be considered in arriving at realistic and equitable solutions to future global problems.

The effectiveness of a regulatory agreement is enhanced when it employs market incentives to stimulate technological innovation. Technology is dynamic and not, as some industrialists have seemed to imply, a static element. But left completely on its own, the market does not necessarily bring forth the right technologies to protect the environment. The ozone protocol set targets that were initially beyond the reach of the existing best-available technology. They appeared difficult but were in fact achievable for most of industry—and thereby averted monolithic industrial opposition that might have delayed international agreement. The treaty actually stimulated collaboration among otherwise competing companies in research and testing, thereby saving both time and money in the development of replacement technologies and products. By expeditiously getting the protocol established in international law even with a 50 percent reduction target, the negotiators effectively signaled the marketplace that research into solutions would now be profitable—thus setting the stage for the later decisions for phaseout.

The signing of a treaty is not necessarily the decisive event in a negotiation; the process before and after signing is critical. It was extremely important to separate the complicated ozone-protection issue into manageable components. The informal fact-finding efforts during 1985 and 1986 and again during 1989 and 1990—workshops, conferences, consultations—established an environment conducive to building personal relationships and generating creative ideas, and thereby facilitated the formal negotiations. During the negotiations themselves, the use of small working groups and a single basic "chairman's text" aided in the gradual emergence of consensus. The developments following the 1987 signing illustrated the wisdom of designing the treaty as a flexible instrument. By providing for periodic integrated assessments—the first of which was advanced from 1990 to 1989 in response to the rapidly changing science—the negotiators made the accord adaptable to evolving circumstances. In effect, the protocol became a continuing process rather than a static solution.

Firmness and pragmatism combined are important ingredients of diplomatic success. The proponents of strong controls refrained from extreme positions but never relented in their pressure for a meaningful treaty. They did not insist on perfect solutions that might have unnecessarily prolonged the negotiations. Nor did they wait for universal participation, or even for agreement among all potential major players. Instead, they achieved an interim solution with built-in flexibility that could serve as a springboard for future action.

Individuals can make a surprisingly significant difference. UNEP's Mostafa Tolba provided overall personal leadership, initiating critical consultations with key governments, private interest groups, and international organizations. During the negotiations, he moved from group to group, arguing for flexibility, applying pressure, often floating his own proposals as a stimulus to the participants. Individual scientists, negotiators, environmentalists, and industry officials also provided ideas, decisions, and actions that proved vital to the final outcome.

Toward Action on Climate Change

The relevance of the experience with the ozone treaties has not been lost on the international community as it addresses greenhouse warming. The establishment in late 1988 and the subsequent functioning of the Intergovernmental Panel on Climate Change—with its multiple scientific, economic, and policy workshops and varied participation from public and private sectors—parallel the fact-gathering and analytical phases of the ozone history. UNEP and the World Meteorological Organization are again actively involved in the process.

Similarly, many governments announced their support during 1989 for a framework agreement on climate change, comparable to the 1985 Vienna Convention on Protecting the Ozone Layer. Such a climate convention need not be a complicated undertaking, and it should be achieved at the earliest possible date. The existence of gaps in scientific and economic knowledge should not become an excuse for delaying the negotiations.

Ideally, a framework convention would enable governments to formalize their agreement in principle on the dimensions of the climate problem and the scope of possible responses. Governments would undertake general obligations for policies and actions to mitigate and adapt to global warming. They would also agree on coordinated research and monitoring to develop additional data as guidelines for future measures.

It might be useful to go beyond the Vienna precedent and try to build into a climate convention some general targets and timetables. However, it would probably be questionable for advocates of stringent greenhouse gas controls to attempt to load a convention with overly detailed and still-controversial commitments. Premature insistence on optimal solutions could have the unintended effect of bogging down the negotiations and prolonging the entire process. On the other hand, an early convention—even if general in its terms—would itself establish a political momentum for further concrete actions.

The framework convention would provide the legal and logistical structure for the crucial step corresponding to the Montreal Protocol: agreement on specific international measures. Work on such measures could well begin even before completion of the framework convention. Because of the complexity of the climate issue, it would not be realistic to attempt to achieve an ideal set of responses at a single stroke; here again, the quest for perfection might only serve to delay action. Instead, it might be more effective—and realistic—to think in terms of incremental stages and partial solutions, following the examples of the Vienna Convention and the Montreal Protocol.

Thus, governments could negotiate a number of separate implementing protocols, each one containing specific measures for dealing with a different aspect of the climate problem. One example would be a treaty mandating greater energy efficiency in the transportation sector, which should be feasible because it need involve only a relative handful of major producers of transportation equipment. The ozone accord itself exemplifies interim progress on the climate problem by means of a constituent protocol. Such partial solutions can be significant; a NASA study estimated that if CFCs had continued to increase at the growth rates of the 1970s, they would by 1989 have surpassed carbon dioxide in their greenhouse impact.[8]

It might be useful to establish standing negotiations under a permanent secretariat, similar to arrangements for the Geneva disarmament talks. By this means, individual protocols could simultaneously be in process of development, each at its own pace.

A climate convention and protocols need not strive right away for universal membership—that could be an unnecessarily complicating factor. In actuality, the overwhelming proportion of carbon emissions from fossil fuels and deforestation is concentrated in a relatively small number of industrialized and developing nations.

Indeed, the major industrialized countries, which are primarily responsible for the world's current precarious ecological condition, could make a vital contribution by agreeing on preemptive actions even before a broader climate treaty is negotiated. The United States and Canada, the Soviet Union, the European Community, and Japan together account for about 60 percent of carbon emissions from fossil fuels.[9] By not delaying actions to increase energy efficiency and reduce carbon dioxide emissions, these countries could significantly slow the warming trend. Doing this would buy time for innovation in energy-efficient technologies and renewable energy sources that could be shared with developing and Eastern European countries to aid them in assuming their own responsibility.

Global Stewardship

Mostafa Tolba has described the Montreal Protocol as "the beginning of a new era of environmental statesmanship."[10] The history of the ozone treaty reflects a new reality: nations must work together in the face of global threats, because if some major actors do not participate, the efforts of others will be vitiated. The process of arriving at the agreement, and the developments that followed its signing, represented new directions for diplomacy, involving unusual emphasis on science and technology, on market forces, on equity, and on flexibility. For all of this, the Montreal Protocol should prove to be a lasting model of international cooperation.

In the realm of international relations, there will always be resistance to change, and there will always be uncertainties—political, economic, scientific, psychological. The ozone protocol's greatest significance, in fact, may be as much in the domain of ethics as environment: it may signal a shift in attitude among critical segments of society in the face of uncertain but potentially grave threats that require coordinated action by sovereign states. The treaty showed that, even in the real world of ambiguity and imperfect knowledge, the international community is capable of undertaking difficult cooperative actions for the benefit of future generations. The Montreal Protocol may thus be the forerunner of an evolving global diplomacy, through which nations accept common responsibility for stewardship of the planet.

Chronology

Publications referred to in the chronology are listed at the end.

Late 1974

American scientists publish theories of CFC-induced ozone layer depletion.

1974–1975

Washington: U.S. congressional hearings on ozone.

1976

First U.S. National Academy of Sciences report.

April 1976

Nairobi: UNEP Governing Council calls for international meeting.

March 1977

Washington: UNEP meeting recommends "World Plan of Action on the Ozone Layer" and establishes Coordinating Committee on the Ozone Layer (CCOL) to produce annual science review.

April 1977

Washington: U.S. hosts first intergovernmental meeting to discuss international regulation of CFCs.

August 1977

Ozone protection amendment to U.S. Clean Air Act.

March 1978

U.S. bans use of CFCs in nonessential aerosols (followed by Canada, Norway, and Sweden).

December 1978

Munich: Federal Republic of Germany hosts second international conference on regulating CFCs.

1979

Major U.S. National Academy of Sciences report.

1980

European Community reduces aerosol use by 30 percent and enacts capacity cap.

April 1980

Nairobi: UNEP Governing Council calls for reductions in CFCs 11 and 12 plus production capacity cap.

May 1981

Nairobi: UNEP Governing Council recommends convention for protection of ozone layer.

1982

Major U.S. National Academy of Sciences report.

January 1982

Stockholm: UNEP convenes Ad Hoc Working Group of Legal and Technical Experts for the Preparation of a Global Framework Convention for the Protection of the Ozone Layer.

1982–1985

Geneva: UNEP Working Group negotiates convention on research, monitoring, and data exchange but fails to agree on protocol on CFC controls.

1984

Major U.S. National Academy of Sciences report.

March 1985

Vienna Convention for the Protection of the Ozone Layer adopted.

May 1985

British scientists publish data showing seasonal Antarctic "ozone hole."

January 1986

NASA ozone assessment released.

EPA releases Stratospheric Ozone Protection Plan outlining domestic and international schedule of meetings and decision points.

February 1986

Nairobi: CCOL meeting on atmospheric science.

May 1986

Washington: U.S. court order settles lawsuit brought by Natural Resources Defense Council against EPA to force unilateral U.S. controls by accepting plans for EPA scientific review and regulatory decision linked to schedule of international negotiations on ozone protection.

Rome: UNEP workshop on CFC trends.

June 1986

Washington: UNEP/EPA international conference and publications on impacts of ozone depletion and climate change on health and environment.

July 1986

WMO/UNEP assessment, Atmospheric Ozone 1985, *published.*

August 1986

U.S. ratifies Vienna Convention.

September 1986

Leesburg, Virginia: UNEP workshop on control strategies.

November 1986

Netherlands: CCOL meeting on effects of ozone depletion.

U.S. negotiating position ("Circular 175") approved.

December 1986

Geneva: first round of protocol negotiations.

United States and Soviet Union agree on joint ozone research.

January 1987

Washington: EPA Science Advisory Board reviews risk assessment study.

January–May 1987

Washington: Senate and House hearings on ozone negotiations.

February 1987

Vienna: second round of protocol negotiations.

February–June 1987

Washington: U.S. Domestic Policy Council reconsiders U.S. negotiating position.

April 1987

Würzburg: UNEP meeting on science models.

Geneva: third round of protocol negotiations.

June 1987

Venice economic summit declaration lists stratospheric ozone depletion first among environmental concerns.

President Reagan formally approves final U.S. position.

Brussels: UNEP informal negotiations among key delegation heads.

July 1987

The Hague: UNEP protocol legal drafting group.

September 1987

Montreal: final round of negotiations—Protocol on Substances That Deplete the Ozone Layer adopted.

January 1988

Washington: trade fair on CFC substitutes.

Paris: meeting of senior advisers to UNEP plans next steps in protocol implementation.

March 1988

Washington: Ozone Trends Panel releases new evidence that CFCs are causing both global ozone depletion and the Antarctic ozone hole.

Du Pont announces phaseout of CFCs.

April 1988

U.S. ratifies Montreal Protocol.

September 1988

Vienna Convention enters into force.

October 1988

The Hague: UNEP conference on science developments, CFC data, legal matters, and substitute products and technologies.

December 1988

Commission of the European Communities together with 8 of 12 member countries ratifies Montreal Protocol. (Other 4 members ratify separately.)

January 1, 1989

Montreal Protocol enters into force.

March 1989

London: Conference on Saving the Ozone Layer.

April–May 1989

Helsinki: First Meeting of Parties to Vienna Convention and Montreal Protocol. Declaration calls for phaseout of CFCs and halons.

August 1989–June 1990

Nairobi and Geneva: meetings of working groups to consider revisions to protocol.

November 1989

UNEP publishes *Synthesis Report* incorporating results of scientific, environmental, economic, and technical assessments.

June 1990

London: Second Meeting of Parties to Montreal Protocol.

Publications (cited in chronological order)

Richard S. Stolarski and Ralph J. Cicerone. "Stratospheric Chlorine: A Possible Sink for Ozone." *Canadian Journal of Chemistry,* no. 52 (1974).

Mario J. Molina and F. Sherwood Rowland. "Stratospheric Sink for Chlorofluoromethanes: Chlorine-Atom Catalyzed Destruction of Ozone." *Nature,* no. 249 (1974).

National Research Council. *Halocarbons: Effects on Stratospheric Ozone.* Washington, D.C.: National Academy Press, 1976.

——— *Stratospheric Ozone Depletion by Halocarbons: Chemistry and Transport.* Washington, D.C.: National Academy Press, 1979.

——— *Causes and Effects of Stratospheric Ozone Reduction: An Update.* Washington, D.C.: National Academy Press, 1982.

——— *Causes and Effects of Changes in Stratospheric Ozone: Update 1983.* Washington, D.C.: National Academy Press, 1984.

J. C. Farman, B. G. Gardiner, and J. D. Shanklin. "Large Losses of Total Ozone in Antarctica Reveal Seasonal Cl_x/NO_x Interaction." *Nature,* no. 315 (1985).

Robert T. Watson, M. A. Geller, Richard S. Stolarski, and R. F. Hampson. *Present State of Knowledge of the Upper Atmosphere.* Washington, D.C.: National Aeronautics and Space Administration, 1986.

EPA. "Stratospheric Ozone Protection Plan." 51 *Fed. Reg.* 1257 (1986).

James G. Titus, ed. *Effects of Changes in Stratospheric Ozone and Global Climate.* Washington, D.C.: EPA, 1986.

WMO. *Atmospheric Ozone 1985: Assessment of Our Understanding of the Processes Controlling Its Present Distribution and Change.* Geneva, 1986.

EPA. "An Assessment of the Risks of Stratospheric Modification." Revised draft, Washington, D.C., January 1987.

UNEP. *Synthesis Report.* UNEP/OzL.Pro.WG.II(1)/4, November 13, 1989.

Appendix A

Vienna Convention for the Protection of the Ozone Layer, March 1985

Preamble

The Parties to this Convention,

Aware of the potentially harmful impact on human health and the environment through modification of the ozone layer,

Recalling the pertinent provisions of the Declaration of the United Nations Conference on the Human Environment, and in particular principle 21, which provides that "States have, in accordance with the Charter of the United Nations and the principles of international law, the sovereign right to exploit their own resources pursuant to their own environmental policies, and the responsibility to ensure that activities within their jurisdiction or control do not cause damage to the environment of other States or of areas beyond the limits of national jurisdiction,"

Taking into account the circumstances and particular requirements of developing countries,

Mindful of the work and studies proceeding within both international and national organizations and, in particular, of the World Plan of Action on the Ozone Layer of the United Nations Environment Programme,

Mindful also of the precautionary measures for the protection of the ozone layer which have already been taken at the national and international levels,

Aware that measures to protect the ozone layer from modifications due to human activities require international co-operation and action, and should be based on relevant scientific and technical considerations,

Aware also of the need for further research and systematic observations to further develop scientific knowledge of the ozone layer and possible adverse effects resulting from its modification,

The text of this appendix is from the Vienna Convention for the Protection of the Ozone Layer, Final Act (Nairobi: UNEP, 1985).

Determined to protect human health and the environment against adverse effects resulting from modifications of the ozone layer,

HAVE AGREED AS FOLLOWS:

Article 1. Definitions

For the purposes of this Convention:

1. "The ozone layer" means the layer of atmospheric ozone above the planetary boundary layer.

2. "Adverse effects" means changes in the physical environment or biota, including changes in climate, which have significant deleterious effects on human health or on the composition, resilience and productivity of natural and managed ecosystems, or on materials useful to mankind.

3. "Alternative technologies or equipment" means technologies or equipment the use of which makes it possible to reduce or effectively eliminate emissions of substances which have or are likely to have adverse effects on the ozone layer.

4. "Alternative substances" means substances which reduce, eliminate or avoid adverse effects on the ozone layer.

5. "Parties" means, unless the text otherwise indicates, Parties to this Convention.

6. "Regional economic integration organization" means an organization constituted by sovereign States of a given region which has competence in respect of matters governed by this Convention or its protocols and has been duly authorized, in accordance with its internal procedures, to sign, ratify, accept, approve or accede to the instruments concerned.

7. "Protocols" means protocols to this Convention.

Article 2. General Obligations

1. The Parties shall take appropriate measures in accordance with the provisions of this Convention and of those protocols in force to which they are party to protect human health and the environment against adverse effects resulting or likely to result from human activities which modify or are likely to modify the ozone layer.

2. To this end the Parties shall, in accordance with the means at their disposal and their capabilities:

(a) Co-operate by means of systematic observations, research and information exchange in order to better understand and assess the effects of human activities on the ozone layer and the effects on human health and the environment from modification of the ozone layer;

(b) Adopt appropriate legislative or administrative measures and co-operate in harmonizing appropriate policies to control, limit, reduce or prevent human activities under their jurisdiction or control should it be found that these activities have or are likely to have adverse effects resulting from modification or likely modification of the ozone layer;

(c) Co-operate in the formulation of agreed measures, procedures and standards for the implementation of this Convention, with a view to the adoption of protocols and annexes;

(d) Co-operate with competent international bodies to implement effectively this Convention and protocols to which they are party.

3. The provisions of this Convention shall in no way affect the right of Parties to adopt, in accordance with international law, domestic measures additional to those referred to in paragraphs 1 and 2 above, nor shall they affect additional domestic measures already taken by a Party, provided that these measures are not incompatible with their obligations under this Convention.

4. The application of this article shall be based on relevant scientific and technical considerations.

Article 3. Research and Systematic Observations

1. The Parties undertake, as appropriate, to initiate and co-operate in, directly or through competent international bodies, the conduct of research and scientific assessments on:

(a) The physical and chemical processes that may affect the ozone layer;

(b) The human health and other biological effects deriving from any modifications of the ozone layer, particularly those resulting from changes in ultra-violet solar radiation having biological effects (UV-B);

(c) Climatic effects deriving from any modifications of the ozone layer;

(d) Effects deriving from any modifications of the ozone layer and any consequent change in UV-B radiation on natural and synthetic materials useful to mankind;

(e) Substances, practices, processes and activities that may affect the ozone layer, and their cumulative effects;

(f) Alternative substances and technologies;

(g) Related socio-economic matters;

and as further elaborated in annexes I and II.

2. The Parties undertake to promote or establish, as appropriate, directly or through competent international bodies and taking fully into account national legislation and relevant ongoing activities at both the national and international levels, joint or complementary programmes for systematic observation of the state of the ozone layer and other relevant parameters, as elaborated in annex I.

3. The Parties undertake to co-operate, directly or through competent international bodies, in ensuring the collection, validation and transmission of research and observational data through appropriate world data centres in a regular and timely fashion.

Article 4. Co-operation in the Legal, Scientific and Technical Fields

1. The Parties shall facilitate and encourage the exchange of scientific, technical, socio-economic, commercial and legal information relevant to this Convention as further elaborated in annex II. Such information shall be supplied to bodies agreed upon by the Parties. Any such body receiving information regarded as confidential by the supplying Party shall ensure that such information is not disclosed and shall aggregate it to protect its confidentiality before it is made available to all Parties.

2. The Parties shall co-operate, consistent with their national laws, regulations and practices and taking into account in particular the needs of the developing countries, in promoting, directly or through competent international bodies, the development and transfer of technology and knowledge. Such co-operation shall be carried out particularly through:

(a) Facilitation of the acquisition of alternative technologies by other Parties;

(b) Provision of information on alternative technologies and equipment, and supply of special manuals or guides to them;

(c) The supply of necessary equipment and facilities for research and systematic observations;

(d) Appropriate training of scientific and technical personnel.

Article 5. Transmission of Information

The Parties shall transmit, through the secretariat, to the Conference of the Parties established under article 6 information on the measures adopted by them in implementation of this Convention and of protocols to which they are party in such form and at such intervals as the meetings of the parties to the relevant instruments may determine.

Article 6. Conference of the Parties

1. A Conference of the Parties is hereby established. The first meeting of the Conference of the Parties shall be convened by the secretariat designated on an interim basis under article 7 not later than one year after entry into force of this Convention. Thereafter, ordinary meetings of the Conference of the Parties shall be held at regular intervals to be determined by the Conference at its first meeting.

2. Extraordinary meetings of the Conference of the Parties shall be held at such other times as may be deemed necessary by the Conference, or at the written

request of any party, provided that, within six months of the request being communicated to them by the secretariat, it is supported by at least one third of the Parties.

3. The Conference of the Parties shall by consensus agree upon and adopt rules of procedure and financial rules for itself and for any subsidiary bodies it may establish, as well as financial provisions governing the functioning of the secretariat.

4. The Conference of the Parties shall keep under continuous review the implementation of this Convention, and, in addition, shall:

(a) Establish the form and the intervals for transmitting the information to be submitted in accordance with article 5 and consider such information as well as reports submitted by any subsidiary body;

(b) Review the scientific information on the ozone layer, on its possible modification and on possible effects of any such modification;

(c) Promote, in accordance with article 2, the harmonization of appropriate policies, strategies and measures for minimizing the release of substances causing or likely to cause modification of the ozone layer, and make recommendations on any other measures relating to this Convention;

(d) Adopt, in accordance with articles 3 and 4, programmes for research, systematic observations, scientific and technological co-operation, the exchange of information and the transfer of technology and knowledge;

(e) Consider and adopt, as required, in accordance with articles 9 and 10, amendments to this Convention and its annexes;

(f) Consider amendments to any protocol, as well as to any annexes thereto, and, if so decided, recommend their adoption to the parties to the protocol concerned;

(g) Consider and adopt, as required, in accordance with article 10, additional annexes to this Convention;

(h) Consider and adopt, as required, protocols in accordance with article 8;

(i) Establish such subsidiary bodies as are deemed necessary for the implementation of this Convention;

(j) Seek, where appropriate, the services of competent international bodies and scientific committees, in particular the World Meteorological Organization and the World Health Organization, as well as the Co-ordinating Committee on the Ozone Layer, in scientific research, systematic observations and other activities pertinent to the objectives of this Convention, and make use as appropriate of information from these bodies and committees;

(k) Consider and undertake any additional action that may be required for the achievement of the purposes of this Convention.

5. The United Nations, its specialized agencies and the International Atomic Energy Agency, as well as any State not party to this Convention, may be represented

at meetings of the Conference of the Parties by observers. Any body or agency, whether national or international, governmental or non-governmental, qualified in fields relating to the protection of the ozone layer which has informed the secretariat of its wish to be represented at a meeting of the Conference of the Parties as an observer may be admitted unless at least one-third of the Parties present object. The admission and participation of observers shall be subject to the rules of procedure adopted by the Conference of the Parties.

Article 7. Secretariat

1. The functions of the secretariat shall be:
   (a) To arrange for and service meetings provided for in articles 6, 8, 9 and 10;
   (b) To prepare and transmit reports based upon information received in accordance with articles 4 and 5, as well as upon information derived from meetings of subsidiary bodies established under article 6;
   (c) To perform the functions assigned to it by any protocols;
   (d) To prepare reports on its activities carried out in implementation of its functions under this Convention and present them to the Conference of the Parties;
   (e) To ensure the necessary co-ordination with other relevant international bodies, and in particular to enter into such administrative and contractual arrangements as may be required for the effective discharge of its functions;
   (f) To perform such other functions as may be determined by the Conference of the Parties.

2. The secretariat functions will be carried out on an interim basis by the United Nations Environment Programme until the completion of the first ordinary meeting of the Conference of the Parties held pursuant to article 6. At its first ordinary meeting, the Conference of the Parties shall designate the secretariat from amongst those existing competent international organizations which have signified their willingness to carry out the secretariat functions under this Convention.

Article 8. Adoption of Protocols

1. The Conference of the Parties may at a meeting adopt protocols pursuant to article 2.

2. The text of any proposed protocol shall be communicated to the Parties by the secretariat at least six months before such a meeting.

Article 9. Amendment of the Convention or Protocols

1. Any Party may propose amendments to this Convention or to any protocol. Such amendments shall take due account, *inter alia,* of relevant scientific and technical considerations.

2. Amendments to this Convention shall be adopted at a meeting of the Conference of the Parties. Amendments to any protocol shall be adopted at a meeting of the Parties to the protocol in question. The text of any proposed amendment to this Convention or to any protocol, except as may otherwise be provided in such protocol, shall be communicated to the Parties by the secretariat at least six months before the meeting at which it is proposed for adoption. The secretariat shall also communicate proposed amendments to the signatories to this Convention for information.

3. The Parties shall make every effort to reach agreement on any proposed amendment to this Convention by consensus. If all efforts at consensus have been exhausted, and no agreement reached, the amendment shall as a last resort be adopted by a three-fourths majority vote of the Parties present and voting at the meeting, and shall be submitted by the Depositary to all Parties for ratification, approval or acceptance.

4. The procedure mentioned in paragraph 3 above shall apply to amendments to any protocol, except that a two-thirds majority of the parties to that protocol present and voting at the meeting shall suffice for their adoption.

5. Ratification, approval or acceptance of amendments shall be notified to the Depositary in writing. Amendments adopted in accordance with paragraph 3 or 4 above shall enter into force between parties having accepted them on the ninetieth day after the receipt by the Depositary of notification of their ratification, approval or acceptance by at least three-fourths of the Parties to this Convention or by at least two-thirds of the parties to the protocol concerned, except as may otherwise be provided in such protocol. Thereafter the amendments shall enter into force for any other Party on the ninetieth day after that Party deposits its instrument of ratification, approval or acceptance of the amendments.

6. For the purposes of this article, "Parties present and voting" means Parties present and casting an affirmative or negative vote.

Article 10. Adoption and Amendment of Annexes

1. The annexes to this Convention or to any protocol shall form an integral part of this Convention or of such protocol, as the case may be, and, unless expressly provided otherwise, a reference to this Convention or its protocols constitutes at the same time a reference to any annexes thereto. Such annexes shall be restricted to scientific, technical and administrative matters.

2. Except as may be otherwise provided in any protocol with respect to its annexes, the following procedure shall apply to the proposal, adoption and entry into force of additional annexes to this Convention or of annexes to a protocol:

(a) Annexes to this Convention shall be proposed and adopted according to the procedure laid down in article 9, paragraphs 2 and 3, while annexes to any protocol shall be proposed and adopted according to the procedure laid down in article 9, paragraphs 2 and 4;

(b) Any party that is unable to approve an additional annex to this Convention or an annex to any protocol to which it is party shall so notify the Depositary, in writing, within six months from the date of the communication of the adoption by the Depositary. The Depositary shall without delay notify all Parties of any such notification received. A Party may at any time substitute an acceptance for a previous declaration of objection and the annexes shall thereupon enter into force for that Party;

(c) On the expiry of six months from the date of the circulation of the communication by the Depositary, the annex shall become effective for all Parties to this Convention or to any protocol concerned which have not submitted a notification in accordance with the provision of subparagraph (b) above.

3. The proposal, adoption and entry into force of amendments to annexes to this Convention or to any protocol shall be subject to the same procedure as for the proposal, adoption and entry into force of annexes to the Convention or annexes to a protocol. Annexes and amendments thereto shall take due account, *inter alia,* of relevant scientific and technical considerations.

4. If an additional annex or an amendment to an annex involves an amendment to this Convention or to any protocol, the additional annex or amended annex shall not enter into force until such time as the amendment to this Convention or to the protocol concerned enters into force.

Article 11. Settlement of Disputes

1. In the event of a dispute between Parties concerning the interpretation or application of this Convention, the parties concerned shall seek solution by negotiation.

2. If the parties concerned cannot reach agreement by negotiation, they may jointly seek the good offices of, or request mediation by, a third party.

3. When ratifying, accepting, approving or acceding to this Convention, or at any time thereafter, a State or regional economic integration organization may declare in writing to the Depositary that for a dispute not resolved in accordance with paragraph 1 or paragraph 2 above, it accepts one or both of the following means of dispute settlement as compulsory:

(a) Arbitration in accordance with procedures to be adopted by the Conference of the Parties at its first ordinary meeting;

(b) Submission of the dispute to the International Court of Justice.

4. If the parties have not, in accordance with paragraph 3 above, accepted the same or any procedure, the dispute shall be submitted to conciliation in accordance with paragraph 5 below unless the parties otherwise agree.

5. A conciliation commission shall be created upon the request of one of the parties to the dispute. The commission shall be composed of an equal number of members appointed by each party concerned and a chairman chosen jointly by the members appointed by each party. The commission shall render a final and recommendatory award, which the parties shall consider in good faith.

6. The provisions of this article shall apply with respect to any protocol except as otherwise provided in the protocol concerned.

Article 12. Signature

This Convention shall be open for signature at the Federal Ministry for Foreign Affairs of the Republic of Austria in Vienna from 22 March 1985 to 21 September 1985, and at United Nations Headquarters in New York from 22 September 1985 to 21 March 1986.

Article 13. Ratification, Acceptance or Approval

1. This Convention and any protocol shall be subject to ratification, acceptance or approval by States and by regional economic integration organizations. Instruments of ratification, acceptance or approval shall be deposited with the Depositary.

2. Any organization referred to in paragraph 1 above which becomes a Party to this Convention or any protocol without any of its member States being a Party shall be bound by all the obligations under the Convention or the protocol as the case may be. In the case of such organizations, one or more of whose member States is a Party to the Convention or relevant protocol, the organization and its member States shall decide on their respective responsibilities for the performance of their obligation under the Convention or protocol, as the case may be. In such cases, the organization and the member States shall not be entitled to exercise rights under the Convention or relevant protocol concurrently.

3. In their instruments of ratification, acceptance or approval, the organizations referred to in paragraph 1 above shall declare the extent of their competence with respect to the matters governed by the Convention or the relevant protocol. These organizations shall also inform the Depositary of any substantial modification in the extent of their competence.

Article 14. Accession

1. This Convention and any protocol shall be open for accession by States and by regional economic integration organizations from the date on which the Convention or the protocol concerned is closed for signature. The instruments of accession shall be deposited with the Depositary.

2. In their instruments of accession, the organizations referred to in paragraph 1 above shall declare the extent of their competence with respect to the matters

governed by the Convention or the relevant protocol. These organizations shall also inform the Depositary of any substantial modification in the extent of their competence.

3. The provisions of article 13, paragraph 2, shall apply to regional economic integration organizations which accede to this Convention or any protocol.

Article 15. Right to Vote

1. Each Party to this Convention or to any protocol shall have one vote.

2. Except as provided for in paragraph 1 above, regional economic integration organizations, in matters within their competence, shall exercise their right to vote with a number of votes equal to the number of their member States which are Parties to the Convention or the relevant protocol. Such organizations shall not exercise their right to vote if their member States exercise theirs, and vice versa.

Article 16. Relationship between the Convention and Its Protocols

1. A State or a regional economic integration organization may not become a party to a protocol unless it is, or becomes at the same time, a Party to the Convention.

2. Decisions concerning any protocol shall be taken only by the parties to the protocol concerned.

Article 17. Entry into Force

1. This Convention shall enter into force on the ninetieth day after the date of deposit of the twentieth instrument of ratification, acceptance, approval or accession.

2. Any protocol, except as otherwise provided in such protocol, shall enter into force on the ninetieth day after the date of deposit of the eleventh instrument of ratification, acceptance or approval of such protocol or accession thereto.

3. For each Party which ratifies, accepts or approves this Convention or accedes thereto after the deposit of the twentieth instrument of ratification, acceptance, approval or accession, it shall enter into force on the ninetieth day after the date of deposit by such Party of its instrument of ratification, acceptance, approval or accession.

4. Any protocol, except as otherwise provided in such protocol, shall enter into force for a party that ratifies, accepts or approves that protocol or accedes thereto after its entry into force pursuant to paragraph 2 above, on the ninetieth day after the date on which that party deposits its instrument of ratification, acceptance, approval or accession, or on the date on which the Convention enters into force for that Party, whichever shall be the later.

5. For the purposes of paragraphs 1 and 2 above, any instrument deposited by a regional economic integration organization shall not be counted as additional to those deposited by member States of such organization.

Article 18. Reservations

No reservations may be made to this Convention.

Article 19. Withdrawal

1. At any time after four years from the date on which this Convention has entered into force for a Party, that Party may withdraw from the Convention by giving written notification to the Depositary.

2. Except as may be provided in any protocol, at any time after four years from the date on which such protocol has entered into force for a party, that party may withdraw from the protocol by giving written notification to the Depositary.

3. Any such withdrawal shall take effect upon expiry of one year after the date of its receipt by the Depositary, or on such later date as may be specified in the notification of the withdrawal.

4. Any Party which withdraws from this Convention shall be considered as also having withdrawn from any protocol to which it is party.

Article 20. Depositary

1. The Secretary-General of the United Nations shall assume the functions of depositary of this Convention and any protocols.

2. The Depositary shall inform the Parties, in particular, of:

(a) The signature of this Convention and of any protocol, and the deposit of instruments of ratification, acceptance, approval or accession in accordance with articles 13 and 14;

(b) The date on which the Convention and any protocol will come into force in accordance with article 17;

(c) Notifications of withdrawal made in accordance with article 19;

(d) Amendments adopted with respect to the Convention and any protocol, their acceptance by the parties and their date of entry into force in accordance with article 9;

(e) All communications relating to the adoption and approval of annexes and to the amendment of annexes in accordance with article 10;

(f) Notifications by regional economic integration organizations of the extent of their competence with respect to matters governed by this Convention and any protocols, and of any modifications thereof.

(g) Declarations made in accordance with article 11, paragraph 3.

Article 21. Authentic Texts

The original of this Convention, of which the Arabic, Chinese, English, French, Russian and Spanish texts are equally authentic, shall be deposited with the Secretary-General of the United Nations.

IN WITNESS WHEREOF the undersigned, being duly authorized to that effect, have signed this Convention.

DONE at Vienna on the 22nd day of March 1985.

Appendix B

Montreal Protocol on Substances That Deplete the Ozone Layer, September 1987

Preamble

The Parties to this Protocol,

Being Parties to the Vienna Convention for the Protection of the Ozone Layer,

Mindful of their obligation under that Convention to take appropriate measures to protect human health and the environment against adverse effects resulting or likely to result from human activities which modify or are likely to modify the ozone layer,

Recognizing that world-wide emissions of certain substances can significantly deplete and otherwise modify the ozone layer in a manner that is likely to result in adverse effects on human health and the environment,

Conscious of the potential climatic effects of emissions of these substances,

Aware that measures taken to protect the ozone layer from depletion should be based on relevant scientific knowledge, taking into account technical and economic considerations,

Determined to protect the ozone layer by taking precautionary measures to control equitably total global emissions of substances that deplete it, with the ultimate objective of their elimination on the basis of developments in scientific knowledge, taking into account technical and economic considerations,

Acknowledging that special provision is required to meet the needs of developing countries for these substances,

Noting the precautionary measures for controlling emissions of certain chlorofluorocarbons that have already been taken at national and regional levels,

Considering the importance of promoting international co-operation in the research and development of science and technology relating to the control and

The text of this appendix is from the Montreal Protocol on Substances That Deplete the Ozone Layer, Final Act (Nairobi: UNEP, 1987).

reduction of emissions of substances that deplete the ozone layer, bearing in mind in particular the needs of developing countries,

HAVE AGREED AS FOLLOWS:

Article 1. Definitions

For the purposes of this Protocol:

1. "Convention" means the Vienna Convention for the Protection of the Ozone Layer, adopted on 22 March 1985.

2. "Parties" means, unless the text otherwise indicates, Parties to this Protocol.

3. "Secretariat" means the secretariat of the Convention.

4. "Controlled substance" means a substance listed in Annex A to this Protocol, whether existing alone or in a mixture. It excludes, however, any such substance or mixture which is in a manufactured product other than a container used for the transportation or storage of the substance listed.

5. "Production" means the amount of controlled substances produced minus the amount destroyed by technologies to be approved by the Parties.

6. "Consumption" means production plus imports minus exports of controlled substances.

7. "Calculated levels" of production, imports, exports and consumption means levels determined in accordance with Article 3.

8. "Industrial rationalization" means the transfer of all or a portion of the calculated level of production of one Party to another, for the purpose of achieving economic efficiencies or responding to anticipated shortfalls in supply as a result of plant closures.

Article 2. Control Measures

1. Each Party shall ensure that for the twelve-month period commencing on the first day of the seventh month following the date of the entry into force of this Protocol, and in each twelve-month period thereafter, its calculated level of consumption of the controlled substances in Group I of Annex A does not exceed its calculated level of consumption in 1986. By the end of the same period, each Party producing one or more of these substances shall ensure that its calculated level of production of the substances does not exceed its calculated level of production in 1986, except that such level may have increased by no more than ten per cent based on the 1986 level. Such increase shall be permitted only so as to satisfy the basic domestic needs of the Parties operating under Article 5 and for the purposes of industrial rationalization between Parties.

2. Each Party shall ensure that for the twelve-month period commencing on the first day of the thirty-seventh month following the date of the entry into force of this Protocol, and in each twelve-month period thereafter, its calculated level of consumption of the controlled substances listed in Group II of Annex A does not exceed its calculated level of consumption in 1986. Each Party producing one or more of these substances shall ensure that its calculated level of production of the substances does not exceed its calculated level of production in 1986, except that such level may have increased by no more than ten per cent based on the 1986 level. Such increase shall be permitted only so as to satisfy the basic domestic needs of the Parties operating under Article 5 and for the purposes of industrial rationalization between Parties. The mechanisms for implementing these measures shall be decided by the Parties at their first meeting following the first scientific review.

3. Each Party shall ensure that for the period 1 July 1993 to 30 June 1994 and in each twelve-month period thereafter, its calculated level of consumption of the controlled substances in Group I of Annex A does not exceed, annually, eighty per cent of its calculated level of consumption in 1986. Each Party producing one or more of these substances shall, for the same periods, ensure that its calculated level of production of the substances does not exceed, annually, eighty per cent of its calculated level of production in 1986. However, in order to satisfy the basic domestic needs of the Parties operating under Article 5 and for the purposes of industrial rationalization between Parties, its calculated level of production may exceed that limit by up to ten per cent of its calculated level of production in 1986.

4. Each Party shall ensure that for the period 1 July 1998 to 30 June 1999, and in each twelve-month period thereafter, its calculated level of consumption of the controlled substances in Group I of Annex A does not exceed, annually, fifty per cent of its calculated level of consumption in 1986. Each Party producing one or more of these substances shall, for the same periods, ensure that its calculated level of production of the substances does not exceed, annually, fifty per cent of its calculated level of production in 1986. However, in order to satisfy the basic domestic needs of the Parties operating under Article 5 and for the purposes of industrial rationalization between Parties, its calculated level of production may exceed that limit by up to fifteen per cent of its calculated level of production in 1986. This paragraph will apply unless the Parties decide otherwise at a meeting by a two-thirds majority of Parties present and voting, representing at least two-thirds of the total calculated level of consumption of these substances of the Parties. This decision shall be considered and made in the light of the assessments referred to in Article 6.

5. Any Party whose calculated level of production in 1986 of the controlled substances in Group I of Annex A was less than twenty-five kilotonnes may, for the purposes of industrial rationalization, transfer to or receive from any other Party, production in excess of the limits set out in paragraphs 1, 3 and 4 provided that the total combined calculated level of production of the Parties concerned

does not exceed the production limits set out in this Article. Any transfer of such production shall be notified to the secretariat, no later than the time of the transfer.

6. Any Party not operating under Article 5, that has facilities for the production of controlled substances under construction, or contracted for, prior to 16 September 1987, and provided for in national legislation prior to 1 January 1987, may add the production from such facilities to its 1986 production of such substances for the purposes of determining its calculated level of production for 1986, provided that such facilities are completed by 31 December 1990 and that such production does not raise that Party's annual calculated level of consumption of the controlled substances above 0.5 kilograms per capita.

7. Any transfer of production pursuant to paragraph 5 or any addition of production pursuant to paragraph 6 shall be notified to the secretariat, no later than the time of the transfer or addition.

8. (a) Any Parties which are Member States of a regional economic integration organization as defined in Article 1(6) of the Convention may agree that they shall jointly fulfil their obligations respecting consumption under this Article provided that their total combined calculated level of consumption does not exceed the levels required by this Article.
   (b) The Parties to any such agreement shall inform the secretariat of the terms of the agreement before the date of the reduction in consumption with which the agreement is concerned.
   (c) Such agreement will become operative only if all Member States of the regional economic integration organization and the organization concerned are Parties to the Protocol and have notified the secretariat of their manner of implementation.

9. (a) Based on the assessments made pursuant to Article 6, the Parties may decide whether:
      (i) adjustments to the ozone depleting potentials specified in Annex A should be made and, if so, what the adjustments should be; and
      (ii) further adjustments and reductions of production or consumption of the controlled substances from 1986 levels should be undertaken and, if so, what the scope, amount and timing of any such adjustments and reductions should be.
   (b) Proposals for such adjustments shall be communicated to the Parties by the secretariat at least six months before the meeting of the Parties at which they are proposed for adoption.
   (c) In taking such decisions, the Parties shall make every effort to reach agreement by consensus. If all efforts at consensus have been exhausted, and no agreement reached, such decisions shall, as a last resort, be adopted by a two-thirds majority vote of the Parties present and voting

representing at least fifty per cent of the total consumption of the controlled substances of the Parties.

(d) The decisions, which shall be binding on all Parties, shall forthwith be communicated to the Parties by the Depositary. Unless otherwise provided in the decisions, they shall enter into force on the expiry of six months from the date of the circulation of the communication by the Depositary.

10. (a) Based on the assessments made pursuant to Article 6 of this Protocol and in accordance with the procedure set out in Article 9 of the Convention, the Parties may decide:
   (i) whether any substances, and if so which, should be added to or removed from any annex to this Protocol; and
   (ii) the mechanism, scope and timing of the control measures that should apply to those substances;

(b) Any such decision shall become effective, provided that it has been accepted by a two-thirds majority vote of the Parties present and voting.

11. Notwithstanding the provisions contained in this Article, Parties may take more stringent measures than those required by this Article.

Article 3. Calculation of Control Levels

For the purposes of Articles 2 and 5, each Party shall, for each Group of substances in Annex A, determine its calculated levels of:

(a) production by:
   (i) multiplying its annual production of each controlled substance by the ozone depleting potential specified in respect of it in Annex A; and
   (ii) adding together, for each such Group, the resulting figures;

(b) imports and exports, respectively, by following, *mutatis mutandis*, the procedure set out in subparagraph (a); and

(c) consumption by adding together its calculated levels of production and imports and subtracting its calculated level of exports as determined in accordance with subparagraphs (a) and (b). However, beginning on 1 January 1993, any export of controlled substances to non-Parties shall not be subtracted in calculating the consumption level of the exporting Party.

Article 4. Control of Trade with Non-Parties

1. Within one year of the entry into force of this Protocol, each Party shall ban the import of controlled substances from any State not party to this Protocol.

2. Beginning on 1 January 1993, no Party operating under paragraph 1 of Article 5 may export any controlled substance to any State not party to this Protocol.

3. Within three years of the date of the entry into force of this Protocol, the Parties shall, following the procedures in Article 10 of the Convention, elaborate in an annex a list of products containing controlled substances. Parties that have not objected to the annex in accordance with those procedures shall ban, within one year of the annex having become effective, the import of those products from any State not party to this Protocol.

4. Within five years of the entry into force of this Protocol, the Parties shall determine the feasibility of banning or restricting, from States not party to this Protocol, the import of products produced with, but not containing, controlled substances. If determined feasible, the Parties shall, following the procedures in Article 10 of the Convention, elaborate in an annex a list of such products. Parties that have not objected to it in accordance with those procedures shall ban or restrict, within one year of the annex having become effective, the import of those products from any State not party to this Protocol.

5. Each Party shall discourage the export, to any State not party to this Protocol, of technology for producing and for utilizing controlled substances.

6. Each Party shall refrain from providing new subsidies, aid, credits, guarantees or insurance programmes for the export to States not party to this Protocol of products, equipment, plants or technology that would facilitate the production of controlled substances.

7. Paragraphs 5 and 6 shall not apply to products, equipment, plants or technology that improve the containment, recovery, recycling or destruction of controlled substances, promote the development of alternative substances, or otherwise contribute to the reduction of emissions of controlled substances.

8. Notwithstanding the provisions of this Article, imports referred to in paragraphs 1, 3 and 4 may be permitted from any State not party to this Protocol if that State is determined, by a meeting of the Parties, to be in full compliance with Article 2 and this Article, and has submitted data to that effect as specified in Article 7.

Article 5. Special Situation of Developing Countries

1. Any Party that is a developing country and whose annual calculated level of consumption of the controlled substances is less than 0.3 kilograms per capita on the date of the entry into force of the Protocol for it, or any time thereafter within ten years of the date of entry into force of the Protocol shall, in order to meet its basic domestic needs, be entitled to delay its compliance with the control measures set out in paragraphs 1 to 4 of Article 2 by ten years after that specified in those paragraphs. However, such Party shall not exceed an annual calculated level of consumption of 0.3 kilograms per capita. Any such Party shall be entitled to use either the average of its annual calculated level of consumption for the period 1995 to 1997 inclusive or a calculated level of consumption of 0.3 kilograms per

capita, whichever is the lower, as the basis for its compliance with the control measures.

2. The Parties undertake to facilitate access to environmentally safe alternative substances and technology for Parties that are developing countries and assist them to make expeditious use of such alternatives.

3. The Parties undertake to facilitate bilaterally or multilaterally the provision of subsidies, aid, credits, guarantees or insurance programmes to Parties that are developing countries for the use of alternative technology and for substitute products.

Article 6. Assessment and Review of Control Measures

Beginning in 1990, and at least every four years thereafter, the Parties shall assess the control measures provided for in Article 2 on the basis of available scientific, environmental, technical and economic information. At least one year before each assessment, the Parties shall convene appropriate panels of experts qualified in the fields mentioned and determine the composition and terms of reference of any such panels. Within one year of being convened, the panels will report their conclusions, through the secretariat, to the Parties.

Article 7. Reporting of Data

1. Each Party shall provide to the secretariat, within three months of becoming a Party, statistical data on its production, imports and exports of each of the controlled substances for the year 1986, or the best possible estimates of such data where actual data are not available.

2. Each Party shall provide statistical data to the secretariat on its annual production (with separate data on amounts destroyed by technologies to be approved by the Parties), imports, and exports to Parties and non-Parties, respectively, of such substances for the year during which it becomes a Party and for each year thereafter. It shall forward the data no later than nine months after the end of the year to which the data relate.

Article 8. Non-Compliance

The Parties, at their first meeting, shall consider and approve procedures and institutional mechanisms for determining non-compliance with the provisions of this Protocol and for treatment of Parties found to be in non-compliance.

Article 9. Research, Development, Public Awareness and Exchange of Information

1. The Parties shall co-operate, consistent with their national laws, regulations and practices and taking into account in particular the needs of developing coun-

tries, in promoting, directly or through competent international bodies, research, development and exchange of information on:

(a) best technologies for improving the containment, recovery, recycling or destruction of controlled substances or otherwise reducing their emissions;

(b) possible alternatives to controlled substances, to products containing such substances, and to products manufactured with them; and

(c) costs and benefits of relevant control strategies.

2. The Parties, individually, jointly or through competent international bodies, shall co-operate in promoting public awareness of the environmental effects of the emissions of controlled substances and other substances that deplete the ozone layer.

3. Within two years of the entry into force of this Protocol and every two years thereafter, each Party shall submit to the secretariat a summary of the activities it has conducted pursuant to this Article.

Article 10. Technical Assistance

1. The Parties shall, in the context of the provisions of Article 4 of the Convention, and taking into account in particular the needs of developing countries, co-operate in promoting technical assistance to facilitate participation in and implementation of this Protocol.

2. Any Party or Signatory to this Protocol may submit a request to the secretariat for technical assistance for the purposes of implementing or participating in the Protocol.

3. The Parties, at their first meeting, shall begin deliberations on the means of fulfilling the obligations set out in Article 9, and paragraphs 1 and 2 of this Article, including the preparation of workplans. Such workplans shall pay special attention to the needs and circumstances of the developing countries. States and regional economic integration organizations not party to the Protocol should be encouraged to participate in activities specified in such workplans.

Article 11. Meetings of the Parties

1. The Parties shall hold meetings at regular intervals. The secretariat shall convene the first meeting of the Parties not later than one year after the date of the entry into force of this Protocol and in conjunction with a meeting of the Conference of the Parties to the Convention, if a meeting of the latter is scheduled within that period.

2. Subsequent ordinary meetings of the Parties shall be held, unless the Parties otherwise decide, in conjunction with meetings of the Conference of the Parties to

the Convention. Extraordinary meetings of the Parties shall be held at such other times as may be deemed necessary by a meeting of the Parties, or at the written request of any Party, provided that, within six months of such a request being communicated to them by the secretariat, it is supported by at least one third of the Parties.

3. The Parties, at their first meeting, shall:
   (a) adopt by consensus rules of procedure for their meetings;
   (b) adopt by consensus the financial rules referred to in paragraph 2 of Article 13;
   (c) establish the panels and determine the terms of reference referred to in Article 6;
   (d) consider and approve the procedures and institutional mechanisms specified in Article 8; and
   (e) begin preparation of workplans pursuant to paragraph 3 of Article 10.

4. The functions of the meetings of the Parties shall be to:
   (a) review the implementation of this Protocol;
   (b) decide on any adjustments or reductions referred to in paragraph 9 of Article 2;
   (c) decide on any addition to, insertion in or removal from any annex of substances and on related control measures in accordance with paragraph 10 of Article 2;
   (d) establish, where necessary, guidelines or procedures for reporting of information as provided for in Article 7 and paragraph 3 of Article 9;
   (e) review requests for technical assistance submitted pursuant to paragraph 2 of Article 10;
   (f) review reports prepared by the secretariat pursuant to subparagraph (c) of Article 12;
   (g) assess, in accordance with Article 6, the control measures provided for in Article 2;
   (h) consider and adopt, as required, proposals for amendment of this Protocol or any annex and for any new annex;
   (i) consider and adopt the budget for implementing this Protocol; and
   (j) consider and undertake any additional action that may be required for the achievement of the purposes of this Protocol.

5. The United Nations, its specialized agencies and the International Atomic Energy Agency, as well as any State not party to this Protocol, may be represented at meetings of the Parties as observers. Any body or agency, whether national or international, governmental or non-governmental, qualified in fields relating to the protection of the ozone layer which has informed the secretariat of its wish to

be represented at a meeting of the Parties as an observer may be admitted unless at least one third of the Parties present object. The admission and participation of observers shall be subject to the rules of procedure adopted by the Parties.

Article 12. Secretariat

For the purposes of this Protocol, the secretariat shall:

(a) arrange for and service meetings of the Parties as provided for in Article 11;

(b) receive and make available, upon request by a Party, data provided pursuant to Article 7;

(c) prepare and distribute regularly to the Parties reports based on information received pursuant to Articles 7 and 9;

(d) notify the Parties of any request for technical assistance received pursuant to Article 10 so as to facilitate the provision of such assistance;

(e) encourage non-Parties to attend the meetings of the Parties as observers and to act in accordance with the provisions of this Protocol;

(f) provide, as appropriate, the information and requests referred to in subparagraphs (c) and (d) to such non-Party observers; and

(g) perform such other functions for the achievement of the purposes of this Protocol as may be assigned to it by the Parties.

Article 13. Financial Provisions

1. The funds required for the operation of this Protocol, including those for the functioning of the secretariat related to this Protocol, shall be charged exclusively against contributions from the Parties.

2. The Parties, at their first meeting, shall adopt by consensus financial rules for the operation of this Protocol.

Article 14. Relationship of This Protocol to the Convention

Except as otherwise provided in this Protocol, the provisions of the Convention relating to its protocols shall apply to this Protocol.

Article 15. Signature

This Protocol shall be open for signature by States and by regional economic integration organizations in Montreal on 16 September 1987, in Ottawa from 17 September 1987 to 16 January 1988, and at United Nations Headquarters in New York from 17 January 1988 to 15 September 1988.

Article 16. Entry into Force

1. This Protocol shall enter into force on 1 January 1989, provided that at least eleven instruments of ratification, acceptance, approval of the Protocol or accession thereto have been deposited by States or regional economic integration organizations representing at least two-thirds of 1986 estimated global consumption of the controlled substances, and the provisions of paragraph 1 of Article 17 of the Convention have been fulfilled. In the event that these conditions have not been fulfilled by that date, the Protocol shall enter into force on the ninetieth day following the date on which the conditions have been fulfilled.

2. For the purposes of paragraph 1, any such instrument deposited by a regional economic integration organization shall not be counted as additional to those deposited by member States of such organization.

3. After the entry into force of this Protocol, any State or regional economic integration organization shall become a Party to it on the ninetieth day following the date of deposit of its instrument of ratification, acceptance, approval or accession.

Article 17. Parties Joining after Entry into Force

Subject to Article 5, any State or regional economic integration organization which becomes a Party to this Protocol after the date of its entry into force, shall fulfil forthwith the sum of the obligations under Article 2, as well as under Article 4, that apply at that date to the States and regional economic integration organizations that became Parties on the date the Protocol entered into force.

Article 18. Reservations

No reservations may be made to this Protocol.

Article 19. Withdrawal

For the purposes of this Protocol, the provisions of Article 19 of the Convention relating to withdrawal shall apply, except with respect to Parties referred to in paragraph 1 of Article 5. Any such Party may withdraw from this Protocol by giving written notification to the Depositary at any time after four years of assuming the obligations specified in paragraphs 1 to 4 of Article 2. Any such withdrawal shall take effect upon expiry of one year after the date of its receipt by the Depositary, or on such later date as may be specified in the notification of the withdrawal.

Article 20. Authentic Texts

The original of this Protocol, of which the Arabic, Chinese, English, French, Russian and Spanish texts are equally authentic, shall be deposited with the Secretary-General of the United Nations.

IN WITNESS WHEREOF the undersigned, being duly authorized to that effect, have signed this protocol.

DONE at Montreal this sixteenth day of September, one thousand nine hundred and eighty seven.

Annex A. Controlled Substances

Group	Substance	Ozone-depleting potential[a]
Group I		
$CFCl_3$	(CFC 11)	1.0
CF_2Cl_2	(CFC 12)	1.0
$C_2F_3Cl_3$	(CFC 113)	0.8
$C_2F_4Cl_2$	(CFC 114)	1.0
C_2F_5Cl	(CFC 115)	0.6
Group II		
CF_2BrCl	halon 1211	3.0
CF_3Br	halon 1301	10.0
$C_2F_4Br_2$	halon 2402	(to be determined)

a. Ozone-depleting potentials are estimates based on existing knowledge and will be reviewed and revised periodically.

Appendix C

London Revisions to the Montreal Protocol, June 1990

Annex I. Adjustments to the Montreal Protocol on Substances That Deplete the Ozone Layer

The Second Meeting of the Parties to the Montreal Protocol on Substances that Deplete the Ozone Layer decides, on the basis of assessments made pursuant to Article 6 of the Protocol, to adopt adjustments and reductions of production and consumption of the controlled substances in Annex A to the Protocol, as follows, with the understanding that:

(a) References in Article 2 to "this Article" and throughout the Protocol to "Article 2" shall be interpreted as references to Articles 2, 2A and 2B;

(b) References throughout the Protocol to "paragraphs 1 to 4 of Article 2" shall be interpreted as references to Articles 2A and 2B; and

(c) The reference in paragraph 5 of Article 2 to "paragraphs 1, 3 and 4" shall be interpreted as a reference to Article 2A.

A. Article 2A: CFCs

Paragraph 1 of Article 2 of the Protocol shall become paragraph 1 of Article 2A, which shall be entitled "Article 2A: CFCs." Paragraphs 3 and 4 of Article 2 shall be replaced by the following paragraphs, which shall be numbered paragraphs 2 to 6 of Article 2A:

2. Each Party shall ensure that for the period from 1 July 1991 to 31 December 1992 its calculated levels of consumption and production of the controlled substances in Group I of Annex A do not exceed 150 per cent of its calculated levels of production and consumption of those substances in 1986; with effect

This text is excerpted from "Report of the Second Meeting of the Parties to the Montreal Protocol on Substances That Deplete the Ozone Layer," UNEP/OzL.Pro.2/3, June 29, 1990. All formal revisions to the protocol, appearing in annexes I and II of the "Report," are included here. In addition, appendixes II and IV of annex IV, annex VII, and the "Statement by Heads of Delegations" are included because of their special importance. (Omitted here are annex III, the noncompliance procedure; and annexes V and VI, budgetary data.)

from 1 January 1993, the twelve-month control period for these controlled substances shall run from 1 January to 31 December each year.

3. Each Party shall ensure that for the twelve-month period commencing on 1 January 1995, and in each twelve-month period thereafter, its calculated level of consumption of the controlled substances in Group I of Annex A does not exceed, annually, fifty per cent of its calculated level of consumption in 1986. Each Party producing one or more of these substances shall, for the same periods, ensure that its calculated level of production of the substances does not exceed, annually, fifty per cent of its calculated level of production in 1986. However, in order to satisfy the basic domestic needs of the Parties operating under paragraph 1 of Article 5, its calculated level of production may exceed that limit by up to ten per cent of its calculated level of production in 1986.

4. Each Party shall ensure that for the twelve-month period commencing on 1 January 1997, and in each twelve-month period thereafter, its calculated level of consumption of the controlled substances in Group I of Annex A does not exceed, annually, fifteen per cent of its calculated level of consumption in 1986. Each Party producing one or more of these substances shall, for the same periods, ensure that its calculated level of production of the substances does not exceed, annually, fifteen per cent of its calculated level of production in 1986. However, in order to satisfy the basic domestic needs of the Parties operating under paragraph 1 of Article 5, its calculated level of production may exceed that limit by up to ten per cent of its calculated level of production in 1986.

5. Each Party shall ensure that for the twelve-month period commencing on 1 January 2000, and in each twelve-month period thereafter, its calculated level of consumption of the controlled substances in Group I of Annex A does not exceed zero. Each Party producing one or more of these substances shall, for the same periods, ensure that its calculated level of production of the substances does not exceed zero. However, in order to satisfy the basic domestic needs of the Parties operating under paragraph 1 of Article 5, its calculated level of production may exceed that limit by up to fifteen per cent of its calculated level of production in 1986.

6. In 1992, the Parties will review the situation with the objective of accelerating the reduction schedule.

B. Article 2B: Halons

Paragraph 2 of Article 2 of the Protocol shall be replaced by the following paragraphs, which shall be numbered paragraphs 1 to 4 of Article 2B:

Article 2B: Halons

1. Each Party shall ensure that for the twelve-month period commencing on 1 January 1992, and in each twelve-month period thereafter, its calculated level of consumption of the controlled substances in Group II of Annex A does not exceed, annually, its calculated level of consumption in 1986. Each Party pro-

ducing one or more of these substances shall, for the same periods, ensure that its calculated level of production of the substances does not exceed, annually, its calculated level of production in 1986. However, in order to satisfy the basic domestic needs of the Parties operating under paragraph 1 of Article 5, its calculated level of production may exceed that limit by up to ten per cent of its calculated level of production in 1986.

2. Each Party shall ensure that for the twelve-month period commencing on 1 January 1995, and in each twelve-month period thereafter, its calculated level of consumption of the controlled substances in Group II of Annex A does not exceed, annually, fifty per cent of its calculated level of consumption in 1986. Each Party producing one or more of these substances shall, for the same periods, ensure that its calculated level of production of the substances does not exceed, annually, fifty per cent of its calculated level of production in 1986. However, in order to satisfy the basic domestic needs of the Parties operating under paragraph 1 of Article 5, its calculated level of production may exceed that limit by up to ten per cent of its calculated level of production in 1986. This paragraph will apply save to the extent that the Parties decide to permit the level of production or consumption that is necessary to satisfy essential uses for which no adequate alternatives are available.

3. Each Party shall ensure that for the twelve-month period commencing on 1 January 2000, and in each twelve-month period thereafter, its calculated level of consumption of the controlled substances in Group II of Annex A does not exceed zero. Each Party producing one or more of these substances shall, for the same periods, ensure that its calculated level of production of the substances does not exceed zero. However, in order to satisfy the basic domestic needs of the Parties operating under paragraph 1 of Article 5, its calculated level of production may exceed that limit by up to fifteen per cent of its calculated level of production in 1986. This paragraph will apply save to the extent that the Parties decide to permit the level of production or consumption that is necessary to satisfy essential uses for which no adequate alternatives are available.

4. By 1 January 1993, the Parties shall adopt a decision identifying essential uses, if any, for the purposes of paragraphs 2 and 3 of this Article. Such decision shall be reviewed by the Parties at their subsequent meetings.

Annex II. Amendment to the Montreal Protocol on Substances That Deplete the Ozone Layer

Article 1. Amendment

A. Preambular Paragraphs

1. The 6th preambular paragraph of the Protocol shall be replaced by the following:

Determined to protect the ozone layer by taking precautionary measures to control equitably total global emissions of substances that deplete it, with the ultimate objective of their elimination on the basis of developments in scientific knowledge, taking into account technical and economic considerations and bearing in mind the developmental needs of developing countries,

2. The 7th preambular paragraph of the Protocol shall be replaced by the following:

Acknowledging that special provision is required to meet the needs of developing countries, including the provision of additional financial resources and access to relevant technologies, bearing in mind that the magnitude of funds necessary is predictable, and the funds can be expected to make a substantial difference in the world's ability to address the scientifically established problem of ozone depletion and its harmful effects,

3. The 9th preambular paragraph of the Protocol shall be replaced by the following:

Considering the importance of promoting international co-operation in the research, development and transfer of alternative technologies relating to the control and reduction of emissions of substances that deplete the ozone layer, bearing in mind in particular the needs of developing countries,

B. Article 1. Definitions

1. Paragraph 4 of Article 1 of the Protocol shall be replaced by the following paragraph:

4. "Controlled substance" means a substance in Annex A or in Annex B to this Protocol, whether existing alone or in a mixture. It includes the isomers of any such substance, except as specified in the relevant Annex, but excludes any controlled substance or mixture which is in a manufactured product other than a container used for the transportation or storage of that substance.

2. Paragraph 5 of Article 1 of the Protocol shall be replaced by the following paragraph:

5. "Production" means the amount of controlled substances produced, minus the amount destroyed by technologies to be approved by the Parties and minus the amount entirely used as feedstock in the manufacture of other chemicals. The amount recycled and reused is not to be considered as "production."

3. The following paragraph shall be added to Article 1 of the Protocol:

9. "Transitional substance" means a substance in Annex C to this Protocol, whether existing alone or in a mixture. It includes the isomers of any such substance, except as may be specified in Annex C, but excludes any transitional substance or mixture which is in a manufactured product other than a container used for the transportation or storage of that substance.

C. Article 2, paragraph 5

Paragraph 5 of Article 2 of the Protocol shall be replaced by the following paragraph:

5. Any Party may, for any one or more control periods, transfer to another Party any portion of its calculated level of production set out in Articles 2A to 2E, provided that the total combined calculated levels of production of the Parties concerned for any group of controlled substances do not exceed the production limits set out in those Articles for that group. Such transfer of production shall be notified to the Secretariat by each of the Parties concerned, stating the terms of such transfer and the period for which it is to apply.

D. Article 2, paragraph 6

The following words shall be inserted in paragraph 6 of Article 2 before the words "controlled substances" the first time they occur:

Annex A or Annex B

E. Article 2, paragraph 8(a)

The following words shall be added after the words "this Article" wherever they appear in paragraph 8(a) of Article 2 of the Protocol:

and Articles 2A to 2E

F. Article 2, paragraph 9(a)(i)

The following words shall be added after "Annex A" in paragraph 9(a)(i) of Article 2 of the Protocol:

and/or Annex B

G. Article 2, paragraph 9(a)(ii)

The following words shall be deleted from paragraph 9(a)(ii) of Article 2 of the Protocol:

from 1986 levels

H. Article 2, paragraph 9(c)

The following words shall be deleted from paragraph 9(c) of Article 2 of the Protocol:

representing at least fifty per cent of the total consumption of the controlled substances of the Parties

and replaced by:

representing a majority of the Parties operating under paragraph 1 of Article 5 present and voting and a majority of the Parties not so operating present and voting

I. Article 2, paragraph 10(b)

Paragraph 10(b) of Article 2 of the Protocol shall be deleted, and paragraph 10(a) of Article 2 shall become paragraph 10.

J. Article 2, paragraph 11

The following words shall be added after the words "this Article" wherever they occur in paragraph 11 of Article 2 of the Protocol:
and Articles 2A to 2E

K. Article 2C. Other Fully Halogenated CFCs

The following paragraphs shall be added to the Protocol as Article 2C:

Article 2C. Other Fully Halogenated CFCs

1. Each Party shall ensure that for the twelve-month period commencing on 1 January 1993, and in each twelve-month period thereafter, its calculated level of consumption of the controlled substances in Group I of Annex B does not exceed, annually, eighty per cent of its calculated level of consumption in 1989. Each Party producing one or more of these substances shall, for the same periods, ensure that its calculated level of production of the substances does not exceed, annually, eighty per cent of its calculated level of production in 1989. However, in order to satisfy the basic domestic needs of the Parties operating under paragraph 1 of Article 5, its calculated level of production may exceed that limit by up to ten per cent of its calculated level of production in 1989.

2. Each Party shall ensure that for the twelve-month period commencing on 1 January 1997, and in each twelve-month period thereafter, its calculated level of consumption of the controlled substances in Group I of Annex B does not exceed, annually, fifteen per cent of its calculated level of consumption in 1989. Each Party producing one or more of these substances shall, for the same periods, ensure that its calculated level of production of the substances does not exceed, annually, fifteen per cent of its calculated level of production in 1989. However, in order to satisfy the basic domestic needs of the Parties operating under paragraph 1 of Article 5, its calculated level of production may exceed that limit by up to ten per cent of its calculated level of production in 1989.

3. Each Party shall ensure that for the twelve-month period commencing on 1 January 2000, and in each twelve-month period thereafter, its calculated level of consumption of the controlled substances in Group I of Annex B does not exceed zero. Each Party producing one or more of these substances shall, for the same periods, ensure that its calculated level of production of the substances does not exceed zero. However, in order to satisfy the basic domestic needs of the Parties operating under paragraph 1 of Article 5, its calculated level of production may exceed that limit by up to fifteen per cent of its calculated level of production in 1989.

L. Article 2D. Carbon Tetrachloride

The following paragraphs shall be added to the Protocol as Article 2D:

Article 2D. Carbon Tetrachloride

1. Each Party shall ensure that for the twelve-month period commencing on 1 January 1995, and in each twelve-month period thereafter, its calculated level of consumption of the controlled substance in Group II of Annex B does not exceed, annually, fifteen per cent of its calculated level of consumption in 1989. Each Party producing the substance shall, for the same periods, ensure that its calculated level of production of the substance does not exceed, annually, fifteen per cent of its calculated level of production in 1989. However, in order to satisfy the basic domestic needs of the Parties operating under paragraph 1 of Article 5, its calculated level of production may exceed that limit by up to ten per cent of its calculated level of production in 1989.

2. Each Party shall ensure that for the twelve-month period commencing on 1 January 2000, and in each twelve-month period thereafter, its calculated level of consumption of the controlled substance in Group II of Annex B does not exceed zero. Each Party producing the substance shall, for the same periods, ensure that its calculated level of production of the substance does not exceed zero. However, in order to satisfy the basic domestic needs of the Parties operating under paragraph 1 of Article 5, its calculated level of production may exceed that limit by up to fifteen per cent of its calculated level of production in 1989.

M. Article 2E. 1,1,1-Trichloroethane (Methyl Chloroform)

The following paragraphs shall be added to the Protocol as Article 2E:

Article 2E. 1,1,1-Trichloroethane (Methyl Chloroform)

1. Each Party shall ensure that for the twelve-month period commencing on 1 January 1993, and in each twelve-month period thereafter, its calculated level of consumption of the controlled substance in Group III of Annex B does not exceed, annually, its calculated level of consumption in 1989. Each Party producing the substance shall, for the same periods, ensure that its calculated level of production of the substance does not exceed, annually, its calculated level of production in 1989. However, in order to satisfy the basic domestic needs of the Parties operating under paragraph 1 of Article 5, its calculated level of production may exceed that limit by up to ten per cent of its calculated level of production in 1989.

2. Each Party shall ensure that for the twelve-month period commencing on 1 January 1995, and in each twelve-month period thereafter, its calculated level of consumption of the controlled substance in Group III of Annex B does not exceed, annually, seventy per cent of its calculated level of consumption in 1989. Each Party producing the substance shall, for the same periods, ensure that its calculated level of production of the substance does not exceed, an-

nually, seventy per cent of its calculated level of consumption in 1989. However, in order to satisfy the basic domestic needs of the Parties operating under paragraph 1 of Article 5, its calculated level of production may exceed that limit by up to ten per cent of its calculated level of production in 1989.

3. Each Party shall ensure that for the twelve-month period commencing on 1 January 2000, and in each twelve-month period thereafter, its calculated level of consumption of the controlled substance in Group III of Annex B does not exceed, annually, thirty per cent of its calculated level of consumption in 1989. Each Party producing the substance shall, for the same periods, ensure that its calculated level of production of the substance does not exceed, annually, thirty per cent of its calculated level of production in 1989. However, in order to satisfy the basic domestic needs of Parties operating under paragraph 1 of Article 5, its calculated level of production may exceed that limit by up to ten per cent of its calculated level of production in 1989.

4. Each Party shall ensure that for the twelve-month period commencing on 1 January 2005, and in each twelve-month period thereafter, its calculated level of consumption of the controlled substance in Group III of Annex B does not exceed zero. Each Party producing the substance shall, for the same periods, ensure that its calculated level of production of the substance does not exceed zero. However, in order to satisfy the basic domestic needs of the Parties operating under paragraph 1 of Article 5, its calculated level of production may exceed that limit by up to fifteen per cent of its calculated level of production in 1989.

5. The Parties shall review, in 1992, the feasibility of a more rapid schedule of reductions than that set out in this Article.

N. Article 3. Calculation of Control Levels

1. The following shall be added after "Articles 2" in Article 3 of the Protocol:
   , 2A to 2E,

2. The following words shall be added after "Annex A" each time it appears in Article 3 of the Protocol:
   or Annex B

O. Article 4. Control of Trade with Non-Parties

1. Paragraphs 1 to 5 of Article 4 shall be replaced by the following paragraphs:

   1. As of 1 January 1990, each Party shall ban the import of the controlled substances in Annex A from any State not party to this Protocol.

   1 *bis*. Within one year of the date of the entry into force of this paragraph, each Party shall ban the import of the controlled substances in Annex B from any State not party to this Protocol.

2. As of 1 January 1993, each Party shall ban the export of any controlled substances in Annex A to any State not party to this Protocol.

2 *bis.* Commencing one year after the date of entry into force of this paragraph, each Party shall ban the export of any controlled substances in Annex B to any State not party to this Protocol.

3. By 1 January 1992, the Parties shall, following the procedures in Article 10 of the Convention, elaborate in an annex a list of products containing controlled substances in Annex A. Parties that have not objected to the annex in accordance with those procedures shall ban, within one year of the annex having become effective, the import of those products from any State not party to this Protocol.

3 *bis.* Within three years of the date of the entry into force of this paragraph, the Parties shall, following the procedures in Article 10 of the Convention, elaborate in an annex a list of products containing controlled substances in Annex B. Parties that have not objected to the annex in accordance with those procedures shall ban, within one year of the annex having become effective, the import of those products from any State not party to this Protocol.

4. By 1 January 1994, the Parties shall determine the feasibility of banning or restricting, from States not party to this Protocol, the import of products produced with, but not containing, controlled substances in Annex A. If determined feasible, the Parties shall, following the procedures in Article 10 of the Convention, elaborate in an annex a list of such products. Parties that have not objected to the annex in accordance with those procedures shall ban, within one year of the annex having become effective, the import of those products from any State not party to this Protocol.

4 *bis.* Within five years of the date of the entry into force of this paragraph, the Parties shall determine the feasibility of banning or restricting, from States not party to this Protocol, the import of products produced with, but not containing, controlled substances in Annex B. If determined feasible, the Parties shall, following the procedures in Article 10 of the Convention, elaborate in an annex a list of such products. Parties that have not objected to the annex in accordance with those procedures shall ban or restrict, within one year of the annex having become effective, the import of those products from any State not party to this Protocol.

5. Each Party undertakes to the fullest practicable extent to discourage the export to any State not party to this Protocol of technology for producing and for utilizing controlled substances.

2. Paragraph 8 of Article 4 of the Protocol shall be replaced by the following paragraph:

8. Notwithstanding the provisions of this Article, imports referred to in paragraphs 1, 1 *bis,* 3, 3 *bis,* 4 and 4 *bis,* and exports referred to in paragraphs 2 and 2 *bis,* may be permitted from, or to, any State not party to this Protocol, if that

State is determined by a meeting of the Parties to be in full compliance with Article 2, Articles 2A to 2E, and this Article and have submitted data to that effect as specified in Article 7.

3. The following paragraph shall be added to Article 4 of the Protocol as paragraph 9:

9. For the purposes of this Article, the term "State not party to this Protocol" shall include, with respect to a particular controlled substance, a State or regional economic integration organization that has not agreed to be bound by the control measures in effect for that substance.

P. Article 5. Special Situation of Developing Countries

Article 5 of the Protocol shall be replaced by the following:

1. Any Party that is a developing country and whose annual calculated level of consumption of the controlled substances in Annex A is less than 0.3 kilograms per capita on the date of the entry into force of the Protocol for it, or any time thereafter until 1 January 1999, shall, in order to meet its basic domestic needs, be entitled to delay for ten years its compliance with the control measures set out in Articles 2A to 2E.

2. However, any Party operating under paragraph 1 of this Article shall exceed neither an annual calculated level of consumption of the controlled substances in Annex A of 0.3 kilograms per capita nor an annual calculated level of consumption of the controlled substances of Annex B of 0.2 kilograms per capita.

3. When implementing the control measures set out in Articles 2A to 2E, any Party operating under paragraph 1 of this Article shall be entitled to use:

(a) For controlled substances under Annex A, either the average of its annual calculated level of consumption for the period 1995 to 1997 inclusive or a calculated level of consumption of 0.3 kilograms per capita, whichever is the lower, as the basis for determining its compliance with the control measures;

(b) For controlled substances under Annex B, the average of its annual calculated level of consumption for the period 1998 to 2000 inclusive or a calculated level of consumption of 0.2 kilograms per capita, whichever is the lower, as the basis for determining its compliance with the control measures.

4. If a Party operating under paragraph 1 of this Article, at any time before the control measures obligations in Articles 2A to 2E become applicable to it, finds itself unable to obtain an adequate supply of controlled substances, it may notify this to the Secretariat. The Secretariat shall forthwith transmit a copy of such notification to the Parties, which shall consider the matter at their next Meeting, and decide upon appropriate action to be taken.

5. Developing the capacity to fulfil the obligations of the Parties operating under paragraph 1 of this Article to comply with the control measures set out in Articles 2A to 2E and their implementation by those same Parties will depend upon the effective implementation of the financial co-operation as provided by Article 10 and transfer of technology as provided by Article 10A.

6. Any Party operating under paragraph 1 of this Article may, at any time, notify the Secretariat in writing that, having taken all practicable steps, it is unable to implement any or all of the obligations laid down in Articles 2A to 2E due to the inadequate implementation of Articles 10 and 10A. The Secretariat shall forthwith transmit a copy of the notification to the Parties, which shall consider the matter at their next Meeting, giving due recognition to paragraph 5 of this Article, and shall decide upon appropriate action to be taken.

7. During the period between notification and the Meeting of the Parties at which the appropriate action referred to in paragraph 6 above is to be decided, or for a further period if the Meeting of the Parties so decides, the non-compliance procedures referred to in Article 8 shall not be invoked against the notifying Party.

8. A Meeting of the Parties shall review, not later than 1995, the situation of the Parties operating under paragraph 1 of this Article, including the effective implementation of financial co-operation and transfer of technology to them, and adopt such revisions as may be deemed necessary regarding the schedule of control measures applicable to those Parties.

9. Decisions of the Parties referred to in paragraphs 4, 6 and 7 of this Article shall be taken according to the same procedure applied to decision-making under Article 10.

Q. Article 6. Assessment and Review of Control Measures

The following words shall be added after "Article 2" in Article 6 of the Protocol:

Articles 2A to 2E, and the situation regarding production, imports and exports of the transitional substances in Group I of Annex C

R. Article 7. Reporting of Data

1. Article 7 of the Protocol shall be replaced by the following:

1. Each Party shall provide to the Secretariat, within three months of becoming a Party, statistical data on its production, imports and exports of each of the controlled substances in Annex A for the year 1986, or the best possible estimates of such data where actual data are not available.

2. Each Party shall provide to the Secretariat statistical data on its production, imports and exports of each of the controlled substances in Annex B and each

of the transitional substances in Group I of Annex C, for the year 1989, or the best possible estimates of such data where actual data are not available, not later than three months after the date when the provisions set out in the Protocol with regard to the substances in Annex B enter into force for that Party.

3. Each Party shall provide statistical data to the Secretariat on its annual production (as defined in paragraph 5 of Article 1) and, separately,

amounts used for feedstocks,

amounts destroyed by technologies approved by the Parties,

imports and exports to Parties and non-Parties respectively,

of each of the controlled substances listed in Annexes A and B as well as of the transitional substances in Group I of Annex C, for the year during which provisions concerning the substances in Annex B entered into force for that Party and for each year thereafter. Data shall be forwarded not later than nine months after the end of the year to which the data relate.

4. For Parties operating under the provisions of paragraph 8(a) of Article 2, the requirements in paragraphs 1, 2 and 3 of this Article in respect of statistical data on imports and exports shall be satisfied if the regional economic integration organization concerned provides data on imports and exports between the organization and States that are not members of that organization.

S. Article 9. Research, Development, Public Awareness and Exchange of Information

Paragraph 1(a) of Article 9 of the Protocol shall be replaced by the following:

(a) Best technologies for improving the containment, recovery, recycling, or destruction of controlled and transitional substances or otherwise reducing their emissions;

T. Article 10. Financial Mechanism

Article 10 of the Protocol shall be replaced by the following:

Article 10. Financial Mechanism

1. The Parties shall establish a mechanism for the purposes of providing financial and technical co-operation, including the transfer of technologies, to Parties operating under paragraph 1 of Article 5 of this Protocol to enable their compliance with the control measures set out in Articles 2A to 2E of the Protocol. The mechanism, contributions to which shall be additional to other financial transfers to Parties operating under that paragraph, shall meet all agreed incremental costs of such Parties in order to enable their compliance with the control measures of the Protocol. An indicative list of the categories of incremental costs shall be decided by the Meeting of the Parties.

2. The mechanism established under paragraph 1 shall include a Multilateral Fund. It may also include other means of multilateral, regional and bilateral co-operation.

3. The Multilateral Fund shall:

(a) Meet, on a grant or concessional basis as appropriate, and according to criteria to be decided upon by the Parties, the agreed incremental costs;

(b) Finance clearing-house functions to:

(i) Assist Parties operating under paragraph 1 of Article 5, through country specific studies and other technical co-operation, to identify their needs for co-operation;

(ii) Facilitate technical co-operation to meet these identified needs;

(iii) Distribute, as provided for in Article 9, information and relevant materials, and hold workshops, training sessions, and other related activities, for the benefit of Parties that are developing countries; and

(iv) Facilitate and monitor other multilateral, regional and bilateral co-operation available to Parties that are developing countries;

(c) Finance the secretarial services of the Multilateral Fund and related support costs.

4. The Multilateral Fund shall operate under the authority of the Parties who shall decide on its overall policies.

5. The Parties shall establish an Executive Committee to develop and monitor the implementation of specific operational policies, guidelines and administrative arrangements, including the disbursement of resources, for the purpose of achieving the objectives of the Multilateral Fund. The Executive Committee shall discharge its tasks and responsibilities, specified in its terms of reference as agreed by the Parties, with the co-operation and assistance of the International Bank for Reconstruction and Development (World Bank), the United Nations Environment Programme, the United Nations Development Programme or other appropriate agencies depending on their respective areas of expertise. The members of the Executive Committee, which shall be selected on the basis of a balanced representation of the Parties operating under paragraph 1 of Article 5 and of the Parties not so operating, shall be endorsed by the Parties.

6. The Multilateral Fund shall be financed by contributions from Parties not operating under paragraph 1 of Article 5 in convertible currency or, in certain circumstances, in kind and/or in national currency, on the basis of the United Nations scale of assessments. Contributions by other Parties shall be encouraged. Bilateral and, in particular cases agreed by a decision of the Parties, regional co-operation may, up to a percentage and consistent with any criteria to be specified by decision of the Parties, be considered as a contribution to the Multilateral Fund, provided that such co-operation, as a minimum:

(a) Strictly relates to compliance with the provisions of this Protocol;

(b) Provides additional resources; and

(c) Meets agreed incremental costs.

7. The Parties shall decide upon the programme budget of the Multilateral Fund for each fiscal period and upon the percentage of contributions of the individual Parties thereto.

8. Resources under the Multilateral Fund shall be disbursed with the concurrence of the beneficiary Party.

9. Decisions by the Parties under this Article shall be taken by consensus whenever possible. If all efforts at consensus have been exhausted and no agreement reached, decisions shall be adopted by a two-thirds majority vote of the Parties present and voting, representing a majority of the Parties operating under paragraph 1 of Article 5 present and voting and a majority of the Parties not so operating present and voting.

10. The financial mechanism set out in this Article is without prejudice to any future arrangements that may be developed with respect to other environmental issues.

U. Article 10A. Transfer of Technology

The following Article shall be added to the Protocol as Article 10A:

Article 10A. Transfer of Technology

Each Party shall take every practicable step, consistent with the programmes supported by the financial mechanism, to ensure:

(a) That the best available, environmentally safe substitutes and related technologies are expeditiously transferred to Parties operating under paragraph 1 of Article 5; and

(b) That the transfers referred to in subparagraph (a) occur under fair and most favourable conditions.

V. Article 11. Meetings of the Parties

Paragraph 4(g) of Article 11 of the Protocol shall be replaced by the following:

(g) Assess, in accordance with Article 6, the control measures and the situation regarding transitional substances;

W. Article 17. Parties Joining after Entry into Force

The following words shall be added after "as well as under" in Article 17:

Articles 2A to 2E, and

X. Article 19. Withdrawal

Article 19 of the Protocol shall be replaced by the following paragraph:

Any Party may withdraw from this Protocol by giving written notification to the Depositary at any time after four years of assuming the obligations specified in paragraph 1 of Article 2A. Any such withdrawal shall take effect upon expiry of one year after the date of its receipt by the Depositary, or on such later date as may be specified in the notification of the withdrawal.

Y. Annexes

The following annexes shall be added to the Protocol:

Annex B. Controlled Substances

Group	Substance	Ozone-depleting potential
Group I		
CF_3Cl	CFC 13	1.0
C_2FCl_5	CFC 111	1.0
$C_2F_2Cl_4$	CFC 112	1.0
C_3FCl_7	CFC 211	1.0
$C_3F_2Cl_6$	CFC 212	1.0
$C_3F_3Cl_5$	CFC 213	1.0
$C_3F_4Cl_4$	CFC 214	1.0
$C_3F_5Cl_3$	CFC 215	1.0
$C_3F_6Cl_2$	CFC 216	1.0
C_3F_7Cl	CFC 217	1.0
Group II		
CCl_4	carbon tetrachloride	1.1
Group III		
$C_2H_3Cl_3$[a]	1,1,1-trichloroethane (methyl chloroform)	0.1

a. This formula does not refer to 1,1,2-trichloroethane.

Annex C. Transitional Substances

Group I	Substance	Group I	Substance
$CHFCl_2$	HCFC 21	$C_3HF_5Cl_2$	HCFC 225
CHF_2Cl	HCFC 22	C_3HF_6Cl	HCFC 226
CH_2FCl	HCFC 31	$C_3H_2FCl_5$	HCFC 231
C_2HFCl_4	HCFC 121	$C_3H_2F_2Cl_4$	HCFC 232
$C_2HF_2Cl_3$	HCFC 122	$C_3H_2F_3Cl_3$	HCFC 233
$C_2HF_3Cl_2$	HCFC 123	$C_3H_2F_4Cl_2$	HCFC 234
C_2HF_4Cl	HCFC 124	$C_3H_2F_5Cl$	HCFC 235
$C_2H_2FCl_3$	HCFC 131	$C_3H_3FCl_4$	HCFC 241
$C_2H_2F_2Cl_2$	HCFC 132	$C_3H_3F_2Cl_3$	HCFC 242
$C_2H_2F_3Cl$	HCFC 133	$C_3H_3F_3Cl_2$	HCFC 243
$C_2H_3FCl_2$	HCFC 141	$C_3H_3F_4Cl$	HCFC 244
$C_2H_3F_2Cl$	HCFC 142	$C_3H_4FCl_3$	HCFC 251
C_2H_4FCl	HCFC 151	$C_3H_4F_2Cl_2$	HCFC 252
C_3HFCl_6	HCFC 221	$C_3H_4F_3Cl$	HCFC 253
$C_3HF_2Cl_5$	HCFC 222	$C_3H_5FCl_2$	HCFC 261
$C_3HF_3Cl_4$	HCFC 223	$C_3H_5F_2Cl$	HCFC 262
$C_3HF_4Cl_3$	HCFC 224	C_3H_6FCl	HCFC 271

Article 2. Entry into Force

1. This Amendment shall enter into force on 1 January 1992, provided that at least twenty instruments of ratification, acceptance or approval of the Amendment have been deposited by States or regional economic integration organizations that are Parties to the Montreal Protocol on Substances that Deplete the Ozone Layer. In the event that this condition has not been fulfilled by that date, the Amendment shall enter into force on the ninetieth day following the date on which it has been fulfilled.

2. For the purposes of paragraph 1, any such instrument deposited by a regional economic integration organization shall not be counted as additional to those deposited by member States of such organization.

3. After the entry into force of this Amendment as provided under paragraph 1, it shall enter into force for any other Party to the Protocol on the ninetieth day following the date of deposit of its instrument of ratification, acceptance or approval.

Additional Material

Annex IV, Appendix II. Terms of Reference of the Executive Committee

1. The Executive Committee of the Parties is established to develop and monitor the implementation of specific operational policies, guidelines and administrative arrangements including the disbursement of resources, for the purpose of achieving the objectives of the Multilateral Fund under the Financial Mechanism.

2. The Executive Committee shall consist of seven Parties from the group of Parties operating under paragraph 1 of Article 5 of the Protocol and seven Parties from the group of Parties not so operating. Each group shall select its Executive Committee members. The members of the Executive Committee shall be formally endorsed by the Meeting of the Parties.

3. The Chairman and Vice-Chairman shall be selected from the fourteen Executive Committee members. The office of Chairman is subject to rotation, on an annual basis, between the Parties operating under paragraph 1 of Article 5, and the Parties not so operating. The group of Parties entitled to the chairmanship shall select the Chairman from among their members of the Executive Committee. The Vice-Chairman shall be selected by the other group from within their number.

4. Decisions by the Executive Committee shall be taken by consensus whenever possible. If all efforts at consensus have been exhausted and no agreement reached, decisions shall be taken by a two-thirds majority of the Parties present and voting, representing a majority of the Parties operating under paragraph 1 of Article 5 and a majority of the Parties not so operating present and voting.

5. The meetings of the Executive Committee shall be conducted in those official languages of the United Nations required by members of the Executive Committee. Nevertheless the Executive Committee may agree to conduct its business in one of the United Nations official languages.

6. Costs of Executive Committee meetings, including travel and subsistence of Committee participants from Parties operating under paragraph 1 of Article 5, shall be disbursed from the Multilateral Fund as necessary.

7. The Executive Committee shall ensure that the expertise required to perform its functions is available to it.

8. The Executive Committee shall meet at least twice a year.

9. The Executive Committee shall adopt other rules of procedure on a provisional basis and in accordance with paragraphs 1 to 8 of these terms of reference. Such provisional rules of procedure shall be submitted to the next annual meeting of the Parties for endorsement. This procedure shall also be followed when such rules of procedure are amended.

10. The functions of the Executive Committee shall include:

(a) To develop and monitor the implementation of specific operational policies, guidelines and administrative arrangements, including the disbursement of resources;

(b) To develop the three-year plan and budget for the Multilateral Fund, including allocation of Multilateral Fund resources among the agencies identified in paragraph 6 of decision II/8;

(c) To supervise and guide the administration of the Multilateral Fund;

(d) To develop the criteria for project eligibility and guidelines for the implementation of activities supported by the Multilateral Fund;

(e) To review regularly the performance reports on the implementation of activities supported by the Multilateral Fund;

(f) To monitor and evaluate expenditure incurred under the Multilateral Fund;

(g) To consider and, where appropriate, approve country programmes for compliance with the Protocol and, in the context of those country programmes, assess and, where applicable, approve all project proposals or groups of project proposals where the agreed incremental costs exceed $500,000;

(h) To review any disagreement by a Party operating under paragraph 1 of Article 5 with any decision taken with regard to a request for financing by that Party of a project or projects where the agreed incremental costs are less than $500,000;

(i) To assess annually whether the contributions through bilateral cooperation, including particular regional cases, comply with the criteria set out by the Parties for consideration as part of the contributions to the Multilateral Fund;

(j) To report annually to the meeting of the Parties on the activities exercised under the functions outlined above, and to make recommendations as appropriate;

(k) To nominate, for appointment by the Executive Director of UNEP, the Chief Officer of the Fund Secretariat, who shall work under the Executive Committee and report to it; and

(l) To perform such other functions as may be assigned to it by the Meeting of the Parties.

Annex IV, Appendix IV. Terms of Reference for the Interim Multilateral Fund

A. Establishment

1. An interim Multilateral Fund of $160 million, which could be raised by up to $80 million during the three-year period when more countries become Parties to

the Protocol, hereinafter referred to as "the Multilateral Fund," shall be established.

B. Roles of the Implementing Agencies

2. Under the overall guidance and supervision of the Executive Committee in the discharge of its policy-making functions:

(a) Implementing agencies shall be requested by the Executive Committee, in the context of country programmes developed to facilitate compliance with the Protocol, to co-operate with and assist the Parties within their respective areas of expertise; and

(b) Implementing agencies shall be invited by the Executive Committee to develop an inter-agency agreement and specific agreements with the Executive Committee acting on behalf of the Parties.

Implementing agencies shall apply only those considerations relevant to effective and economically efficient programmes and projects which are consistent with any criteria adopted by the Parties.

3. Specifically,

(a) The United Nations Environment Programme shall be invited by the Executive Committee to co-operate and assist in political promotion of the objectives of the Protocol, as well as in research, data gathering and the clearing-house functions;

(b) The United Nations Development Programme and such other agencies which, within their areas of expertise, may be able to assist, shall be invited by the Executive Committee to co-operate and assist in feasibility and pre-investment studies and in other technical assistance measures;

(c) The World Bank shall be invited by the Executive Committee to co-operate and assist in administering and managing the programme to finance the agreed incremental costs;

(d) Other agencies, in particular regional development banks, shall also be invited by the Executive Committee to co-operate with and assist it in carrying out its functions.

4. The Executive Committee shall draw up reporting criteria and shall invite the implementing agencies to report regularly to it in accordance with those criteria.

5. The Executive Committee shall invite the implementing agencies, in fulfilling their responsibilities in respect of the Multilateral Fund, to consult each other regularly. It shall also invite the heads of the agencies, or their representatives, to meet at least once a year to report on their activities and consult on co-operative arrangements.

6. The implementing agencies shall be entitled to receive support costs for the activities they undertake having reached specific agreements with the Executive Committee.

C. Budget and Contributions

7. The Multilateral Fund shall be financed in accordance with paragraph 7 of decision II/8. In addition, contributions may be made by countries not Party to the Protocol, and by other governmental, intergovernmental, non-governmental and other sources.

8. The contributions referred to in paragraph 7 above are to be based on the scale of contributions set out in Appendix III. Bilateral, and in particular cases, regional co-operation by a country not operating under paragraph 1 of Article 5 may, according to criteria adopted by the Parties, be considered as a contribution to the Multilateral Fund up to a total of twenty per cent of the total contribution by that Party set out in Appendix III.

9. All contributions other than the value of bilateral and agreed regional co-operation referred to in paragraph 8 above shall be in convertible currency or, in certain circumstances, in kind and/or in national currency.

10. Contributions from States that become Parties not operating under paragraph 1 of Article 5 after the beginning of the financial period of the mechanism shall be calculated on a *pro rata* basis for the balance of the financial period.

11. Contributions not immediately required for the purposes of the Multilateral Fund shall be invested under the authority of the Executive Committee and any interest so earned shall be credited to the Multilateral Fund.

12. Budget estimates, setting out the income and expenditure of the Multilateral Fund prepared in United States dollars, shall be drawn up by the Executive Committee and submitted to the regular meetings of the Parties to the Protocol.

13. The proposed budget estimates shall be dispatched by the Fund Secretariat to all Parties to the Protocol at least sixty days before the date fixed for the opening of the regular meeting of the Parties to the Protocol at which they are to be considered.

14. After entry into force of the Amendment to the Protocol, the Financial Mechanism shall be established by the Parties at their next regular meeting and any resources remaining in the interim Multilateral Fund shall be transferred to the multilateral fund established under that mechanism.

D. Administration

15. The World Bank shall be invited by the Executive Committee to co-operate with and assist it in administering and managing the programme to finance the agreed incremental costs of Parties operating under paragraph 1 of Article 5. Should the World Bank accept this invitation, in the context of an agreement with the Executive Committee, the President of the World Bank shall be the Administrator of this programme, which shall operate under the authority of the Executive Committee.

16. The Executive Committee shall encourage the involvement of other agencies, in particular the regional development banks, in carrying out its functions effectively in relation to the programme to finance the agreed incremental costs.

17. The Fund Secretariat operating under the Chief Officer, co-located with the United Nations Environment Programme (UNEP) at a place to be decided by the Executive Committee, shall assist the Executive Committee in the discharge of its functions. The Multilateral Fund shall cover Secretariat costs, based on regular budgets to be submitted for decision by the Executive Committee.

18. In the event that the Chief Officer of the Fund Secretariat anticipates that there may be a shortfall in resources over the financial period as a whole, he shall have discretion to adjust the budget approved by the Parties so that expenditures are at all times fully covered by contributions received.

19. No commitments shall be made in advance of the receipt of contributions, but income not spent in a budget year and unimplemented activities may be carried forward from one year to the next within the financial period.

20. At the end of each calendar year, the Chief Officer of the Fund Secretariat shall submit to the Parties accounts for the year. The Chief Officer shall also, as soon as practicable, submit the audited accounts for each period so as to coincide with the accounting procedures of the implementing agencies.

21. The Fund Secretariat and the implementing agencies shall co-operate with the Parties to provide information on funding available for relevant projects, to secure the necessary contacts and to co-ordinate, when requested by the interested Party, projects financed from other sources with activities financed under the Protocol.

22. The financing of activities or other costs, including resources channelled to third party beneficiaries, shall require the concurrence of the recipient Governments concerned. Recipient Governments shall, where appropriate, be associated with the planning of projects and programmes.

23. Nothing shall preclude a beneficiary Party operating under paragraph 1 of Article 5 from applying for its requirements for agreed incremental costs solely from the resources available to the Multilateral Fund.

Annex VII. Resolution by the Governments and the European Communities Represented at the Second Meeting of the Parties to the Montreal Protocol

The Governments and the European Communities represented at the Second Meeting of the Parties to the Montreal Protocol
Resolve:

I. Other Halons Not Listed in Annex A, Group II, of the Montreal Protocol ("Other Halons")

1. To refrain from authorizing or to prohibit production and consumption of fully halogenated compounds containing one, two or three carbon atoms and at least one atom each of bromine and fluorine,[1] and not listed in Group II of Annex A of the Montreal Protocol (hereafter called "other halons"), which are of such a chemical nature or such a quantity that they would pose a threat to the ozone layer;

2. To refrain from using other halons except for those essential applications where other more environmentally suitable alternative substances or technologies are not yet available; and

3. To report to the Secretariat to the Protocol estimates of their annual production and consumption of such other halons;

II. Transitional Substances

1. To apply the following guidelines to facilitate the adoption of transitional substances with a low ozone-depleting potential, such as hydrochlorofluorocarbons (HCFCs), where necessary, and their timely substitution by non-ozone depleting and more environmentally suitable alternative substances or technologies:

(a) Use of transitional substances should be limited to those applications where other more environmentally suitable alternative substances or technologies are not available;

(b) Use of transitional substances should not be outside the areas of application currently met by the controlled and transitional substances, except in rare cases for the protection of human life or human health;

(c) Transitional substances should be selected in a manner that minimizes ozone depletion, in addition to meeting other environmental, safety and economic considerations;

(d) Emission control systems, recovery and recycling should, to the degree possible, be employed in order to minimize emissions to the atmosphere;

(e) Transitional substances should, to the degree possible, be collected and prudently destroyed at the end of their final use;

2. To review regularly the use of transitional substances, their contribution to ozone depletion and global warming, and the availability of alternative products and application technologies, with a view to their replacement by non-ozone depleting and more environmentally suitable alternatives and as the scientific evi-

1. The list of other halons will appear in the *Montreal Protocol Handbook*, to be prepared by the Executive Director.

dence requires: at present, this should be no later than 2040 and, if possible, no later than 2020;

III. 1,1,1-Trichloroethane (Methyl Chloroform)

1. To phase out production and consumption of methyl chloroform as soon as possible;

2. To request the Technology Review Panel to investigate the earliest technically feasible dates for reductions and total phase-out; and

3. To request the Technology Review Panel to report their findings to the preparatory meeting of the Parties with a view to the consideration by the Meeting of the Parties, not later than 1992;

IV. More Stringent Measures

1. To express appreciation to those Parties that have already taken measures more stringent and broader in scope than those required by the Protocol;

2. To urge adoption, in accordance with the spirit of paragraph 11 of Article 2 of the Protocol, of such measures in order to protect the ozone layer.

Statement by Heads of Delegations of the Governments of Australia, Austria, Belgium, Canada, Denmark, Federal Republic of Germany, Finland, Liechtenstein, Netherlands, New Zealand, Norway, Sweden, Switzerland

The heads of delegations of the above governments represented at the Second Meeting of the Parties to the Montreal Protocol

Concerned about the recent scientific findings on severe depletion of the ozone layer of both Southern and Northern Hemispheres;

Mindful that all CFCs are also powerful greenhouse gases leading to global warming,

Convinced of the availability of more environmentally suitable alternative substances or technologies, and

Convinced of the need to further tighten control measures of CFCs beyond the Protocol adjustments agreed by the Parties to the Montreal Protocol,

DECLARE their firm determination to take all appropriate measures to phase out the production and consumption of all fully halogenated chlorofluorocarbons controlled by the Montreal Protocol, as adjusted and amended, as soon as possible but not later than 1997.

Appendix D

Countries Signing and Ratifying the Vienna Convention and the Montreal Protocol as of August 2, 1990

Signature of a treaty by a country does not by itself entail a legal obligation, but rather an intention to ratify or accede to the treaty. Ratification or accession makes the country legally a party to the treaty. A country can accede without having previously signed the treaty.

The symbol + indicates that the country is an original signatory: March 22, 1985, for the Vienna Convention; September 16, 1987, for the Montreal Protocol.

	Vienna Convention		Montreal Protocol	
Country	Signed	Ratified	Signed	Ratified
Afghanistan				
Albania				
Algeria				
Angola				
Antigua and Barbuda				
Argentina	3/85+	1/90	6/88	
Australia		9/87	6/88	5/89
Austria	9/85	8/87	8/88	5/89
Bahamas				
Bahrain		4/90		4/90
Bangladesh		8/90		8/90
Barbados				
Belgium	3/85+	10/88	9/87+	12/88
Belize				
Benin				
Bhutan				
Bolivia				
Botswana				
Brazil		3/90		3/90
Brunei Darussalam		7/90		
Bulgaria				
Burkina Faso	12/85	3/89	9/88	7/89

	Vienna Convention		Montreal Protocol	
Country	Signed	Ratified	Signed	Ratified
Burundi				
Byelorussian S.S.R.	3/85+	6/86	1/88	10/88
Cambodia				
Cameroon		8/89		8/89
Canada	3/85+	6/86	9/87+	6/88
Cape Verde				
Central African Republic				
Chad		5/89		
Chile	3/85+	3/90	6/88	3/90
China		9/89		
Colombia		7/90		
Comoros				
Congo			9/88	
Costa Rica				
Côte d'Ivoire				
Cuba				
Cyprus				
Czechoslovakia				
Denmark	3/85+	9/88	9/87+	12/88
Djibouti				
Dominica				
Dominican Republic				
Ecuador		4/90		4/90
Egypt	3/85+	5/88	9/87+	8/88
El Salvador				
Equatorial Guinea		8/88		
Ethiopia				
Fiji		10/89		10/89
Finland	3/85+	9/86	9/87+	12/88
France	3/85+	12/87	9/87+	12/88
Gabon				
Gambia		7/90		7/90
Germany, Dem. Rep.		1/89		1/89
Germany, Fed. Rep.	3/85+	9/88	9/87+	12/88
Ghana		7/89	9/87+	7/89
Greece	3/85+	12/88	10/87	12/88
Grenada				
Guatemala		9/87		11/89

Country	Vienna Convention		Montreal Protocol	
	Signed	Ratified	Signed	Ratified
Guinea				
Guinea-Bissau				
Guyana				
Haiti				
Holy See				
Honduras				
Hungary		5/88		4/89
Iceland		8/89		8/89
India				
Indonesia			7/88	
Iran				
Iraq				
Ireland		9/88	9/88	12/88
Israel			1/88	
Italy	3/85+	9/88	9/87+	12/88
Jamaica				
Japan		9/88	9/87+	9/88
Jordan		5/89		5/89
Kenya		11/88	9/87+	11/88
Kiribati				
Korea, North				
Korea, South				
Kuwait				
Laos				
Lebanon				
Lesotho				
Liberia				
Libya		7/90		7/90
Liechtenstein		2/89		2/89
Luxembourg	4/85	10/88	1/88	10/88
Madagascar				
Malawi				
Malaysia		8/89		8/89
Maldives		4/88	7/88	5/89
Mali				
Malta		9/88	9/88	12/88
Mauritania				
Mauritius				

Country	Vienna Convention		Montreal Protocol	
	Signed	Ratified	Signed	Ratified
Mexico	4/85	9/87	9/87+	3/88
Mongolia				
Morocco	2/86		1/88	
Mozambique				
Myanmar (Burma)				
Nauru				
Nepal				
Netherlands	3/85+	9/88	9/87+	12/88
New Zealand	3/86	6/87	9/87+	7/88
Nicaragua				
Niger				
Nigeria		10/88		10/88
Norway	3/85+	9/86	9/87+	6/88
Oman				
Pakistan				
Panama		2/89	9/87+	3/89
Papua New Guinea				
Paraguay				
Peru	3/85+	4/89		
Philippines			9/88	
Poland		7/90		7/90
Portugal		10/88	9/87+	10/88
Qatar				
Romania				
Rwanda				
Saint Christopher and Nevis				
Saint Lucia				
Saint Vincent and the Grenadines				
Samoa (Western)				
Sao Tome and Principe				
Saudi Arabia				
Senegal			9/87+	
Seychelles				
Sierra Leone				
Singapore		1/89		1/89
Solomon Islands				

Country	Vienna Convention		Montreal Protocol	
	Signed	Ratified	Signed	Ratified
Somalia				
South Africa		1/90		1/90
Spain		7/88	7/88	12/88
Sri Lanka		12/89		12/89
Sudan				
Suriname				
Swaziland				
Sweden	3/85+	11/86	9/87+	6/88
Switzerland	3/85+	12/87	9/87+	12/88
Syria		12/89		12/89
Tanzania				
Thailand		7/89	9/88	7/89
Togo			9/87+	
Tonga				
Trinidad and Tobago		8/89		8/89
Tunisia		9/89		9/89
Turkey				
Tuvalu				
Uganda		6/88	9/88	9/88
Ukrainian S.S.R.	3/85+	6/86	2/88	9/88
U.S.S.R.	3/85+	6/86	12/87	11/88
United Arab Emirates		12/89		12/89
United Kingdom	5/85	5/87	9/87+	12/88
United States	3/85+	8/86	9/87+	4/88
Uruguay		2/89		
Vanuatu				
Venezuela		9/88	9/87+	2/89
Vietnam				
Yemen				
Yemen, South				
Yugoslavia		4/90		
Zaire				
Zambia		1/90		1/90
Zimbabwe				
European Economic Community	3/85+	10/88	9/87+	12/88

Notes

1. Lessons from History

1. Lee M. Thomas, testimony before U.S. Senate, Committee on Foreign Relations, Hearings, February 19, 1988, in *Ozone Protocol* (Washington, D.C.: U.S. Government Printing Office, 1988), p. 61; Ronald Reagan, statement, April 5, 1988, reprinted in "President Signs Protocol on Ozone-Depleting Substances," *Department of State Bulletin*, June 1988, p. 30; George Mitchell, statement in U.S. Senate, *Ozone Protocol*, p. 53.
2. For the United States alone, with under 30 percent of world production, the market value of CFCs produced in 1985 was about $750 million, the value of goods and services in industries directly dependent on CFCs was approximately $27 billion, related employment numbered over 700,000, and installed equipment and products requiring CFCs for maintenance and repair were estimated at $135 billion. See Alliance for Responsible CFC Policy, *An Economic Portrait of the CFC-Utilizing Industries of the United States* (Rosslyn, Va., 1985), pp. 10–11.
3. U.K. House of Lords, *Hansard* 500 (October 20, 1988): col. 1308.
4. *Declaration of the United Nations Conference on the Human Environment* (1972; reprint, Nairobi: UNEP).
5. See, e.g., Richard E. Benedick, "The Environment on the Foreign Policy Agenda," *Ecology Law Quarterly* 13, no. 2 (1986):173–174; Ernst-Ulrich Freiherr von Weizsäcker, "Umweltschutz: Eine neue Dimension der internationalen Politik," in *Nach Vorn Gedacht . . . Perspektiven Deutscher Aussenpolitik*, ed. H.-D. Gentscher (Stuttgart: Bonn Aktuell, 1987); Organisation for Economic Cooperation and Development, *State of Environment 1985* (Paris, 1985), pp. 244–246, which contains a list of 50 international environmental treaties; UNEP, *Register of International Treaties and Other Agreements in the Field of the Environment* (Nairobi, 1985), which lists 118 such agreements from 1921 through 1983.
6. Richard E. Benedick, "International Efforts to Protect the Stratospheric Ozone Layer," address before plenary meeting of UNEP, Vienna, February 23, 1987, in U.S. Department of State, Bureau of Public Affairs, *Current Policy No. 931* (Washington, D.C., 1987).
7. Mostafa K. Tolba, "The Tools to Build a Global Response," statement to meeting of Working Group of Parties to the Montreal Protocol, Nairobi, August 21, 1989, UNEP press release.
8. Reagan, statement, April 5, 1988, p. 30.

2. The Science: Models of Uncertainty

1. Guy Brasseur, "The Endangered Ozone Layer," *Environment* 29, no. 1 (1987): 8; Susan Solomon, "The Mystery of the Antarctic Ozone Hole," *Reviews of Geophysics* 26, no. 1 (1988):131; R. D. Bojkov, "Surface Ozone during the Second Half of the Nineteenth Century," *Journal of Climate and Applied Meteorology* 25, no. 3 (1986):344–345.
2. Daniel L. Albritton et al., *Stratospheric Ozone: The State of the Science and NOAA's Current and Future Research* (Washington, D.C.: National Oceanic and Atmospheric Administration, 1987), p. 1.
3. UNEP, *The Ozone Layer* (Nairobi, 1987), pp. 8–9; Guy Brasseur and Paul Simon, "Changes in Stratospheric Ozone: Observations and Theories," *Aeronomica Acta*, A-No.334-1988 (Brussels: Institut d'Aeronomie Spatiale de Belgique, 1988), p. 8; James G. Titus, ed., *Effects of Changes in Stratospheric Ozone and Global Climate* (Washington, D.C.: EPA, 1986); and EPA, "An Assessment of the Risks of Stratospheric Modification," Revised draft, Washington, D.C., January 1987.
4. UNEP, *The Ozone Layer*, p. 10; Robert T. Watson, M. A. Geller, Richard S. Stolarski, and R. F. Hampson, *Present State of Knowledge of the Upper Atmosphere* (Washington, D.C.: National Aeronautics and Space Administration, 1986), chap. 12.
5. Richard S. Stolarski and Ralph J. Cicerone, "Stratospheric Chlorine: A Possible Sink for Ozone," *Canadian Journal of Chemistry* 52 (1974):1610–15.
6. Mario J. Molina and F. Sherwood Rowland, "Stratospheric Sink for Chlorofluoromethanes: Chlorine Atomic Catalysed Destruction of Ozone," *Nature* 249 (1974):810–812.
7. Chemical Manufacturers Association, "Production, Sales, and Calculated Release of CFC-11 and CFC-12 through 1987," Washington, D.C., 1988.
8. Lydia Dotto and Harold Schiff, *The Ozone War* (Garden City, N.Y.: Doubleday, 1978), chaps. 1, 3, 11.
9. See, e.g., the following publications by the National Research Council: *Halocarbons: Effects on Stratospheric Ozone* (Washington, D.C.: National Academy Press, 1976); *Stratospheric Ozone Depletion by Halocarbons: Chemistry and Transport* (Washington, D.C.: National Academy Press, 1979); *Causes and Effects of Stratospheric Ozone Reduction: An Update* (Washington, D.C.: National Academy Press, 1982); *Causes and Effects of Changes in Stratospheric Ozone: Update 1983* (Washington, D.C.: National Academy Press, 1984).
10. F. Sherwood Rowland, "Chlorofluorocarbons and the Depletion of Stratospheric Ozone," *American Scientist* 77 (January–February 1989):37–39, 41–43; Brasseur and Simon, "Changes in Stratospheric Ozone," pp. 9–11.
11. Dotto and Schiff, *The Ozone War*, pp. 149–165.
12. R. L. McCarthy, testimony before the U.S. House of Representatives, Committee on Interstate and Foreign Commerce, Hearings, December 11–12, 1974, in *Fluorocarbons: Impact on Health and Environment* (Washington, D.C.: U.S. Government Printing Office, 1974), p. 381.
13. D. J. Wuebbles, F. M. Luther, and J. E. Penner, "Effect of Coupled Anthropogenic Perturbations on Stratospheric Ozone," *Journal of Geophysical Research* 88 (1983):1444; Brasseur and Simon, "Changes in Stratospheric Ozone," pp. 11–15.

14. WMO, *Atmospheric Ozone 1985: Assessment of Our Understanding of the Processes Controlling Its Present Distribution and Change* (Geneva, 1986), p. 2.
15. Guy Brasseur and Susan Solomon, *Aeronomy of the Middle Atmosphere* (Hingham, Mass.: D. Reidel, 1984), p. 410; National Research Council, *Changes in Stratospheric Ozone: Update 1983,* p. 101.
16. WMO, *Atmospheric Ozone 1985,* p. 4.
17. Ibid., chap. 13.
18. Ibid., pp. 786–787.
19. Ibid., chaps. 9, 13.
20. Ibid., chap. 15.
21. Ibid., chap. 12.
22. Joseph Scotto et al., "Biologically Effective Ultraviolet Radiation: Surface Measurements in the United States, 1974–85," *Science* 239 (1988):762–764.
23. WMO, *Atmospheric Ozone 1985,* pp. 8, 14, 819.
24. Watson et al., *Present State of Knowledge,* p. xiii.
25. Alan S. Miller and Irving M. Mintzer, *The Sky Is the Limit* (Washington, D.C.: World Resources Institute, 1986), p. 24.
26. Albritton et al., *Stratospheric Ozone,* p. 22.
27. Rowland, "Chlorofluorocarbons and Depletion of Stratospheric Ozone," p. 43.
28. J. C. Farman, B. G. Gardiner, and J. D. Shanklin, "Large Losses of Total Ozone in Antarctica Reveal Seasonal ClO_x / NO_x Interaction," *Nature* 315 (1985):207–210.
29. Solomon, "Mystery of Antarctic Ozone Hole," p. 133; John Gribbin, *The Hole in the Sky* (New York: Bantam, 1988), p. 111.
30. Personal discussions with F. Sherwood Rowland, Robert T. Watson, and Ralph J. Cicerone.
31. WMO, *Atmospheric Ozone 1985,* p. 791.
32. See, e.g., Albritton et al., *Stratospheric Ozone,* p. ix; WMO, *Atmospheric Ozone 1985,* chap. 14; Watson et al., *Present State of Knowledge,* p. 15.
33. Solomon, "Mystery of Antarctic Ozone Hole," pp. 144–146; Richard S. Stolarski, "The Antarctic Ozone Hole," *Scientific American* 258, no. 1 (January 1988):35–36; Boyce Rensberger, "New Theory for Polar Ozone Hole," *Washington Post,* May 25, 1987; Brasseur and Simon, "Changes in Stratospheric Ozone," pp. 23–33.
34. F. Sherwood Rowland, "A Threat to Earth's Protective Shield," *EPA Journal* 12, no. 10 (1986):4.
35. F. Sherwood Rowland, "Can We Close the Ozone Hole?" *Technology Review,* August/September 1987, p. 51.
36. Peter M. Morrisette, "The Evolution of Policy Responses to Stratospheric Ozone Depletion," *Natural Resources Journal* 29 (1989):814–815.
37. EPA, "Protection of Stratospheric Ozone," 52 *Fed. Reg.* 47492 (1987).
38. Titus, *Effects of Changes;* EPA, "Assessment of Risks."
39. Margaret L. Kripke, letter to U.S. Representative John L. Dingell, April 7, 1987, in U.S. House of Representatives, Committee on Energy and Commerce, Hearings, March 9, 1987, *Ozone Layer Depletion* (Washington, D.C.: U.S. Government Printing Office, 1987), pp. 616–622 (the discrepancy in dates reflects the fact that documents were submitted after the hearings); E. M. Sloss and T. P. Rose, *Possible*

Health Effects of Increased Exposure to Ultraviolet Radiation (Santa Monica, Calif.: Rand Corporation, 1985), pp. 1–70; EPA, "Assessment of Risks," chaps. 7 and 8.

40. Kripke, letter to Dingell.
41. EPA, "Protection of Stratospheric Ozone," p. 47494; EPA, "Assessment of Risks," chap. 7.
42. EPA, "Protection of Stratospheric Ozone," p. 47495; National Research Council, *Changes in Stratospheric Ozone: Update 1983*, chap. 8.
43. EPA, "Protection of Stratospheric Ozone," p. 47495; Alan Teramura, "Overview of Our Current State of Knowledge of UV Effects on Plants," in Titus, *Effects of Changes*, pp. 165–173.
44. EPA, "Protection of Stratospheric Ozone," p. 47495; R. C. Worrest, "The Effect of Solar UV-B Radiation on Aquatic Systems: An Overview," in Titus, *Effects of Changes*, pp. 175–191.
45. WMO, *Atmospheric Ozone 1985*, chap. 15; UNEP, Coordinating Committee on the Ozone Layer, "UNEP Policy Support Document," UNEP/WG.151/Background 4, December 1, 1986 (unless specified otherwise, all UNEP documents are deposited at UNEP headquarters in Nairobi); WMO and UNEP, "Developing Policies for Responding to Climatic Change," in *World Climate Programme Impact Studies* (Geneva, 1988).

3. Spray Cans and Europolitics

1. The European Community consists of Belgium, Denmark, France, the Federal Republic of Germany, Greece, Ireland, Luxembourg, the Netherlands, Portugal, Spain and the United Kingdom. The EC Commission, with headquarters in Brussels, is one part of a dual executive; it has responsibility for initiating and implementing EC policy. The other part of the executive, the Council of Ministers, is composed of ministerial representatives of the individual EC member governments and is the main decision-making body. Ministers responsible for different issues may be represented on the Council depending on the subject under consideration.
2. Clean Air Act, 42 U.S.C. §7457(b).
3. EPA, "Protection of Stratospheric Ozone," 52 *Fed. Reg.* 47491 (1987).
4. Toxic Substances Control Act, 15 U.S.C. §2605, sec. 6; 43 *Fed. Reg.* 11301–19 (1978).
5. Lydia Dotto and Harold Schiff, *The Ozone War* (Garden City, N.Y.: Doubleday, 1978), p. 148.
6. See discussion in Markus Jachtenfuchs, "The European Community and the Protection of the Ozone Layer," *Journal of Common Market Studies* 28, no. 3 (March 1980):263.
7. Council of Europe, Parliamentary Assembly, Committee on Science and Technology, Hearings, *Prohibition of the Use of Chlorofluorocarbons and Other Measures to Preserve the Ozone Layer* (Strasbourg, 1980), p. 21. The Council of Europe is an organization of European governments founded after World War II, broader in membership than the EC and more general in its objectives of consultation and cooperation.

8. *Schutz der Erdatmosphäre: Eine internationale Herausforderung,* Zwischenbericht der Enquete-Kommission des 11. Deutschen Bundestages (Bonn, 1988), p. 200.
9. Nigel Haigh, *EEC Environmental Policy and Britain,* 2d ed. (London: Longman, 1989), p. 266.
10. EC, "Council Decision of 26 March 1980 concerning Chlorofluorocarbons in the Environment," 80/372/EEC, and "Council Decision of 15 November 1982 on the Consolidation of Precautionary Measures concerning Chlorofluorocarbons in the Environment," 82/795/EEC, Brussels.
11. David W. Pearce, "The European Community Approach to the Control of Chlorofluorocarbons" (Paper submitted to UNEP Workshop on the Control of Chlorofluorocarbons, Leesburg, Va., September 8–12, 1986), p. 12.
12. Jachtenfuchs, "European Community and Protection," p. 263.
13. For European Community CFC production and sales see Pearce, "European Community Approach to Control," pp. 16, 18, derived from data of the European Fluorocarbon Technical Committee of the European Chemical Industry Federation. For German actions, see *Schutz der Erdatmosphäre,* p. 203.
14. Haigh, *EEC Environmental Policy and Britain,* p. 268. See also Sylvie Faucheux and J.-F. Noël, *Did the Ozone War End in Montreal?* Université de Paris, Centre Economie-Espace-Environnement, English digest (Paris: Cahiers du C.3.E., 1988), p. 5; Jachtenfuchs, "European Community and Protection," p. 263.
15. Pearce, "European Approach to Control," p. 19.
16. Haigh, *EEC Environmental Policy and Britain,* pp. 267–268.
17. Council of Europe, *Prohibition of Use of Chlorofluorocarbons,* p. 55; *Schutz der Erdatmosphäre,* pp. 196, 202.
18. Data in this section, unless otherwise noted, refer only to CFCs 11 and 12 and not to CFC 113, for which comparable data were not available. Moreover, because data for some producers were unavailable or incomplete, the U.S. and EC proportions are slightly (perhaps 5 percent) overstated as given. Sources include: Pearce, "European Community Approach to Control," p. 16; Chemical Manufacturers Association, "Production, Sales, and Calculated Release of CFC-11 and CFC-12 through 1987," Washington, D.C., 1988; and ICF Incorporated, "Chlorofluorocarbon Production and Use Data," Report to EPA, Washington, D.C., February 1987.
19. Organisation for Economic Cooperation and Development, *Impact of Restrictions on the Use of Fluorocarbons, Final Report* (Paris, 1978), p. 12. Spain at this time was not yet a member of the European Community.
20. EPA estimates, based on national reports and other sources, itemized in U.S. House of Representatives, Committee on Energy and Commerce, Hearings, March 9, 1987, *Ozone Layer Depletion* (Washington, D.C.: U.S. Government Printing Office, 1987), pp. 406–415.
21. U.S. Congress, Office of Technology Assessment, "An Analysis of the Montreal Protocol on Substances That Deplete the Ozone Layer," Revised staff paper, Washington, D.C., February 1988, p. 11.
22. "Comments of 'Eurimpact' (a CFC user group) to the Commission of the European Communities on Implementation of the Montreal Protocol," Brussels, November 9, 1987, p. 10; Faucheux and Noël, *Did the Ozone War End in Montreal?* p. 6.

23. Faucheux and Noël, *Did the Ozone War End in Montreal?* p. 7.
24. M. D. Lemonick, "The Heat Is On," *Time,* October 19, 1987, pp. 58–63; R. H. Boyle, "Forecast for Disaster," *Sports Illustrated,* November 16, 1987, pp. 78–84.
25. See, e.g., Thomas B. Stoel, Jr., Alan S. Miller, and Breck Milroy, *Fluorocarbon Regulation* (Lexington, Mass.: D. C. Heath, 1980), p. 205; J. T. B. Tripp, D. J. Dudek, and Michael Oppenheimer, "Equity and Ozone Protection," *Environment* 29, no. 6 (1987):45.
26. Examples of such studies include Stoel, Miller, and Milroy, *Fluorocarbon Regulation;* and Alan S. Miller and Irving M. Mintzer, *The Sky Is the Limit* (Washington, D.C.: World Resources Institute, 1986).
27. Natural Resources Defense Council v. Thomas 84-3587 (D.D.C. 1985).
28. U.S. House of Representatives, Committee on Interstate and Foreign Commerce, Hearings, December 11–12, 1974, *Fluorocarbons: Impact on Health and Environment* (Washington, D.C.: U.S. Government Printing Office, 1974); U.S. Senate, Committee on Aeronautical and Space Sciences, Hearings, September 1975, *Stratospheric Ozone Depletion* (Washington, D.C.: U.S. Government Printing Office, 1975).
29. U.S. House of Representatives, Concurrent Resolution 47, February 18, 1987, and Concurrent Resolution 50, February 19, 1987; U.S. Senate, Concurrent Resolution 19, February 19, 1987, and S.226, June 5, 1987.
30. Richard E. Benedick, "International Efforts to Protect the Stratospheric Ozone Layer," address before plenary meeting of UNEP, Vienna, February 23, 1987, in U.S. Department of State, Bureau of Public Affairs, *Current Policy No. 931* (Washington, D.C., 1987). Bills introduced in the House of Representatives included H.R. 2036 and H.R. 2854; in the Senate, S.570 and S.571.
31. Hearings were held, for example, by the U.S. Senate Committee on Environment and Public Works and by the U.S. House of Representatives Committees on Foreign Affairs, on Energy and Commerce, and on Science, Space and Technology.
32. Dotto and Schiff, *The Ozone War,* chap. 7.
33. "Role of European Science on the Stratospheric Ozone Problem," statement issued in Snowmass, Colo., May 1988 (author's files); Christine McGourty, "Britain's Ozone Research Lags a Long Way behind the Rest," *Nature* 335 (1988):657.
34. John Gribbin, *The Hole in the Sky* (New York: Bantam, 1988), p. 164.
35. Alexis de Tocqueville, *Democracy in America,* vol. 2 (New York: Alfred A. Knopf, 1951), p. 123.
36. R. L. Schuyler, testimony in U.S. Senate, *Stratospheric Ozone Depletion,* p. 570.
37. Anthony Lewis, "The Long and the Short," *New York Times,* May 11, 1989.
38. M. S. Weil, "Chlorofluorocarbon Update: Stir Caused by EPA-Supported World Ban Proposal," *Contracting Business,* October 1984, pp. 11–12; Alliance for Responsible CFC Policy, *The Montreal Protocol: A Briefing Book* (Rosslyn, Va., 1987), pp. v, I-1,2.
39. Alliance for Responsible CFC Policy, *The Montreal Protocol,* pp. I-2,3. See also "Du Pont Position Statement on the Chlorofluorocarbon/Ozone/Greenhouse Issues," *Environmental Conservation* 13, no. 4 (1986):363–364.
40. Alliance for Responsible CFC Policy, *The Montreal Protocol,* p. I-3.
41. Faucheux and Noël, *Did the Ozone War End in Montreal?* p. 7, and personal observations of author.

42. Dotto and Schiff, *The Ozone War,* p. 116.
43. Gribbin, *The Hole in the Sky,* p. 168.
44. Paul Brodeur, "Annals of Chemistry: In the Face of Doubt," *New Yorker,* June 9, 1986, p. 87; E. I. du Pont de Nemours and Co., "Freon" Products Division, "Fluorocarbon/Ozone Update," Wilmington, Del., March 1987.
45. Alan S. Miller, "Incentives for CFC Substitutes: Lessons for Other Greenhouse Gases," in *Coping with Climate Change,* ed. J. C. Topping, Jr. (Washington, D.C.: Climate Institute, 1989), pp. 548–549.
46. For further discussion of industry responses to the protocol see Chapters 8, 10, and 11.
47. Winfried Lang, "Diplomatie zwischen Ökonomie und Ökologie: Das Beispiel des Ozonvertrags von Montreal," *Europa Archiv: Zeitschrift für Internationale Politik,* February 2, 1988, p. 108; Jachtenfuchs, "European Community and Protection," p. 263.
48. *Schutz der Erdatmosphäre,* p. 206.
49. G. Diprose and D. W. Reddy, "An Industry Perspective on the Chlorofluoromethane/Ozone Issue," testimony in Council of Europe, *Prohibition of Use of Chlorofluorocarbons,* p. 37. See also European Chemical Industry Federation, "Critique of the U.S. EPA's Papers," June 1985, cited in Miller and Mintzer, *The Sky Is the Limit,* p. 24.
50. Peter Winsemius, "The Two-Speed Europe," *European Affairs* (1987):76; Beate Weber, "Die ungeliebte Gemeinschaft: Über den Umgang mit europäischer Umweltpolitik," in *Dicke Luft in Europa: Aufgaben und Probleme der europäischen Umweltpolitik,* ed. Lothar Gundling and Beate Weber (Heidelberg: Müller, 1988), pp. 3–20.
51. Faucheux and Noël, *Did the Ozone War End in Montreal?* p. 7.
52. Jachtenfuchs, "European Community and Protection," p. 275.
53. Ibid., p. 262. See also Haigh, *EEC Environmental Policy and Britain,* p. 363; U.K. House of Lords, Session 1987–88, Select Committee on the European Communities, "Minutes of Evidence, 21 June 1988," in *The Ozone Layer: Implementing the Montreal Protocol* (London: Her Majesty's Stationery Office, 1988), pp. 7, 11; "Special Treatment for the European Community" in Chapter 7.
54. U.K. House of Lords, "Seventeenth Report, 12 July 1988," in *The Ozone Layer,* p. 11.
55. Carlo Ripa di Meana, press conference, Brussels, February 7, 1989 (author's files).
56. The Conference Board, *1992: Leading Issues for European Companies,* Research Report no. 921 (New York and Brussels, 1989), p. 17.
57. Ripa di Meana, press conference.
58. Jachtenfuchs, "European Community and Protection," pp. 264, 273.
59. Ernst-Ulrich Freiherr von Weizsäcker, "Umweltschutz: Eine neue Dimension der internationalen Politik," in *Nach Vorn Gedacht . . . Perspektiven Deutscher Aussenpolitik,* ed. H.-D. Gentscher (Stuttgart: Bonn Aktuell, 1987), p. 199 (translation by author).
60. *Schutz der Erdatmosphäre,* pp. 199, 203, 208.
61. Faucheux and Noël, *Did the Ozone War End in Montreal?* p. 7.
62. Stoel, Miller, and Milroy, *Fluorocarbon Regulation,* pp. 250–251.

63. Jachtenfuchs, "European Community and Protection," p. 265.
64. Cited in ibid.
65. L. J. Brinkhorst, "Ozone: Europe Is Moving," *International Herald Tribune*, June 21, 1988.
66. Jachtenfuchs, "European Community and Protection," p. 274.
67. S. Clinton Davis, speech to the Ozone Depletion Conference, Royal Institute of British Architects, London, November 28, 1988, p. 7.
68. Richard E. Benedick, "U.S. Environmental Policy: Relevance to Europe," *International Environmental Affairs* 1, no. 2 (1989):91–93.
69. Jachtenfuchs, "European Community and Protection," p. 275.
70. Stoel, Miller, and Milroy, *Fluorocarbon Regulation*, p. 204.
71. Comments to author and other members of U.S. delegation. See also Gribbin, *The Hole in the Sky*, pp. 168–169; Jachtenfuchs, "European Community and Protection," p. 263.
72. See, e.g., Council of Europe, *Prohibition of Use of Chlorofluorocarbons*, p. 18; U.K. Department of the Environment, "Chlorofluorocarbons and Their Effect on Stratospheric Ozone," Pollution Paper no. 5, London, 1976.
73. U.K. Department of the Environment, "Chlorofluorocarbons and Their Effect," pp. 16, 9.
74. Council of Europe, *Prohibition of Use of Chlorofluorocarbons*, p. 42.
75. Dotto and Schiff, *The Ozone War*, pp. 24, 110–111, 156–157.
76. U.K. House of Lords, "Seventeenth Report," p. 6.
77. Jachtenfuchs, "European Community and Protection," pp. 265, 268; also compare U.K. government and industry reports cited in Stoel, Miller, and Milroy, *Fluorocarbon Regulation*, pp. 63–64.
78. Cable in author's files.
79. *Schutz der Erdatmosphäre*, pp. 208–211.

4. Prelude to Consensus

1. For a detailed account of UNEP activities, see Peter S. Thacher, "Stratospheric Ozone Depletion," background note for Center for the Study of Global Habitability, Conference on Science, Public Policy, and Climate Change, National Academy of Sciences, Washington, D.C., May 3, 1988.
2. WMO, "Statement on Modification of the Ozone Layer Due to Human Activities and Some Possible Geophysical Consequences," WMO/R/STW/2, annex, Geneva, 1975.
3. UNEP, "Report of the UNEP Meeting of Experts Designated by Governments, Intergovernmental and Nongovernmental Organizations on the Ozone Layer," UNEP/WG/7/25/Rev.1, annex 3, March 8, 1977.
4. *Schutz der Erdatmosphäre: Eine internationale Herausforderung*, Zwischenbericht der Enquete-Kommission des 11. Deutschen Bundestages (Bonn, 1988), p. 195.
5. Thomas B. Stoel, Jr., Alan S. Miller, and Breck Milroy, *Fluorocarbon Regulation* (Lexington, Mass.: D. C. Heath, 1980), p. 275.
6. Ibid., p. 52; *Schutz der Erdatmosphäre*, p. 195.

7. Iwona Rummel-Bulska, "Recent Developments Relating to the Vienna Convention for the Protection of the Ozone Layer," *Yearbook of the AAA* (Association of Attenders and Alumni), vol. 54/55/56 (Dordrecht: Martinus Nijhoff, 1986), p. 117.
8. Sylvie Faucheux and J.-F. Noël, *Did the Ozone War End in Montreal?* Université de Paris, Centre Economie-Espace-Environnement, English digest (Paris: Cahiers du C.3.E., 1988), p. 7.
9. Markus Jachtenfuchs, "The European Community and the Protection of the Ozone Layer," *Journal of Common Market Studies* 28, no. 3 (March 1990):263.
10. U.S. Department of State, *Protecting the Ozone Layer,* Public Information Series (Washington, D.C., 1985).
11. Iain Guest, "U.S. and E.C. Split on Danger to Ozone," *International Herald Tribune,* January 29, 1985.
12. Johan G. Lammers, "Efforts to Develop a Protocol on Chlorofluorocarbons to the Vienna Convention for the Protection of the Ozone Layer," *The Hague Yearbook of International Law* 1 (1988):227–228.
13. Patrick Szell, "The Vienna Convention for the Protection of the Ozone Layer," *International Digest of Health Legislation* 36, no. 3 (1985):840, 841.
14. Winfried Lang, "Environmental Protection," *Journal of World Trade Law* 20, no. 5 (1986):492.
15. UNEP, Vienna Convention for the Protection of the Ozone Layer, Final Act (Nairobi: UNEP, 1985), annex I, p. 29.
16. Ibid., resolution 2, pp. 7–8.
17. Rummel-Bulska, "Recent Developments," pp. 122–123.
18. Malone was assistant secretary for oceans and international environmental and scientific affairs in the Department of State and had earlier played a key role for the administration in opposing the Law of the Sea Treaty. On the ozone protection issue, however, he worked hard for an international convention because of the pragmatic case for more and better research.
19. Chemical Manufacturers Association, "Production, Sales, and Calculated Release of CFC-11 and CFC-12 through 1985," Washington, D.C., 1986.
20. William E. Mooz, Kathleen A. Wolf, and Frank Camm, *Potential Constraints on Cumulative Global Production of Chlorofluorocarbons* (Santa Monica, Calif.: Rand Corporation, 1986).
21. Estimates by E. I. du Pont de Nemours and Co. See also *Schutz der Erdatmosphäre,* p. 209.
22. EPA, "Stratospheric Ozone Protection Plan," 51 *Fed. Reg.* 1257 (1986); WMO, *Atmospheric Ozone 1985: Assessment of Our Understanding of the Processes Controlling Its Present Distribution and Change* (Geneva, 1986); James G. Titus, ed., *Effects of Changes in Stratospheric Ozone and Global Climate* (Washington, D.C.: EPA, 1986).
23. "Some UNEP Workshop Delegates Say Protocol on Ozone Protection Possible by Spring 1987," *International Environment Reporter* 9, no. 10 (October 1986): 346–348.
24. UNEP, "Report of the Second Part of the Workshop on the Control of Chlorofluorocarbons," Leesburg, Va., September 8–12, 1986, UNEP/WG.148/3, annex II, p. 3.

5. Forging the U.S. Position

1. Clean Air Act, 42 U.S.C. §7456.
2. "Du Pont Position Statement on the Chlorofluorocarbon/Ozone/Greenhouse Issues," *Environmental Conservation* 13, no. 4 (1986):363–364. See also Alan S. Miller and Irving M. Mintzer, *The Sky Is the Limit* (Washington, D.C.: World Resources Institute, 1986), p. 16.
3. "Circular 175: Request for Authority to Negotiate a Protocol to the Convention for the Protection of the Ozone Layer," memorandum from Assistant Secretary of State John Negroponte to Under Secretary of State Allen Wallis, November 28, 1986, reprinted in U.S. House of Representatives, Committee on Energy and Commerce, Hearings, March 9, 1987, *Ozone Layer Depletion* (Washington, D.C.: U.S. Government Printing Office, 1987), pp. 119–129.
4. Ibid., p. 121.
5. For further details on the U.S. position as it evolved during the negotiations, see U.S. Senate, Committee on Environment and Public Works, Joint Hearings, January 28, 1987, *Ozone Depletion, the Greenhouse Effect, and Climate Change* (Washington, D.C.: U.S. Government Printing Office, 1987), pp. 44–54, 135–200; U.S. Senate, Committee on Environment and Public Works, Joint Hearings, May 12–14, 1987, *Stratospheric Ozone Depletion and Chlorofluorocarbons* (Washington, D.C.: U.S. Government Printing Office, 1987), pp. 475–496; U.S. House of Representatives, *Ozone Layer Depletion,* pp. 93–162; U.S. House of Representatives, Committee on Science, Space, and Technology, Hearings, March 10 and 12, 1987, *Stratospheric Ozone Depletion* (Washington, D.C.: U.S. Government Printing Office, 1987), pp. 153–178; and U.S. House of Representatives, Committee on Foreign Affairs, Hearings, March 5, 1987, *U.S. Participation in International Negotiations on Ozone Protocol* (Washington, D.C.: U.S. Government Printing Office, 1987).
6. Letters to author from Charles Wick, director, U.S. Information Agency.
7. "Economic Declaration, Venice Economic Summit," *Department of State Bulletin* 87, no. 2125 (August 1987):14.
8. Sharon Roan, *Ozone Crisis* (New York: John Wiley & Sons, 1989), p. 195.
9. U.S. Senate, *Ozone Depletion, the Greenhouse Effect, and Climate Change* and *Stratospheric Ozone Depletion and Chlorofluorocarbons;* U.S. House of Representatives, *Ozone Layer Depletion, Stratospheric Ozone Depletion,* and *U.S. Participation in International Negotiations.*
10. U.S. House of Representatives, *Ozone Layer Depletion,* pp. 6, 7.
11. John L. Dingell, letter to Secretary of the Interior Donald Hodel, June 1, 1987, reprinted in *Congressional Record: House,* June 29, 1987, pp. H5726, H5727.
12. Margaret L. Kripke, letter to David Gibbons, deputy associate director, U.S. Office of Management and Budget, April 6, 1987, reprinted in U.S. House of Representatives, *Ozone Layer Depletion,* pp. 623–625.
13. R. E. Taylor, "Advice on Ozone May Be: 'Wear Hats and Stand in Shade'—Aides Hint U.S. Won't Move to Stop Chemical Erosion of the Ultraviolet Barrier," *Wall Street Journal,* May 29, 1987.
14. Cass Peterson, "Administration Ozone Policy May Favor Sunglasses, Hats," *Washington Post,* May 29, 1987.

15. "D. Hodel, Boy Environmentalist," editorial, *Washington Post,* June 7, 1987; Herblock cartoon, *Washington Post,* June 3, 1987; Rochelle L. Stanfield, "New Black Hat," *National Journal,* June 20, 1987; "Through Rose-Colored Sunglasses," editorial, *New York Times,* May 31, 1987.
16. *Congressional Record: Senate,* June 4, 1987, pp. S7610, S7658; ibid., June 5, 1987, p. S7717; *Congressional Record: House,* June 29, 1987, p. H5721.
17. Dingell, letter to Hodel.
18. Donald Hodel, letter to Senator Timothy Wirth, June 4, 1987, reprinted in *Congressional Record: Senate,* June 5, 1987, pp. S7708–09; Donald Hodel, letter to editor, *Energy Daily,* July 6, 1987.
19. Senate Resolution 226, "Relating to International Negotiations to Protect the Ozone Layer," *Congressional Record: Senate,* June 5, 1987, p. S7759.
20. Hodel, letter to Wirth.
21. *Congressional Record: Senate,* June 5, 1987, p. S7712.
22. Ibid.
23. Representative Mike Synar, letter to Donald Hodel, September 10, 1987 (author's files); see also *Congressional Record: House,* June 29, 1987, pp. H5721–22.
24. "President Must Decide—State Department Pushes Radical Ozone Treaty," *Human Events,* June 20, 1987. See also *World Environment Report* 13, no. 15 (1987):113.
25. Philip Shabecoff, "A Wrangle over Ozone Policy," *New York Times,* June 23, 1987.
26. Whitehead was acting secretary while George Shultz was traveling on missions abroad, including accompanying President Reagan to the 1987 economic summit in Venice, which raised the priority of ozone layer protection.
27. Senate Resolution 226.
28. Stanfield, "New Black Hat."

6. The Sequence of Negotiations

1. Imperial Chemical Industries, "Chlorofluorocarbons and the Ozone Layer," brochure, Runcorn, Cheshire, 1986, p. 1.
2. Personal communication to the author.
3. Commission of the European Communities, "Communication from the Commission to the Council [of Ministers]: Chlorofluorocarbons in the Environment: A Reexamination of Control Measures; Proposal for a Council Decision," COM(86) 602 final, Brussels, November 20, 1986.
4. Markus Jachtenfuchs, "The European Community and the Protection of the Ozone Layer," *Journal of Common Market Studies* 28, no. 3 (March 1990):265.
5. *Schutz der Erdatmosphäre: Eine internationale Herausforderung,* Zwischenbericht der Enquete-Kommission des 11. Deutschen Bundestages (Bonn, 1988), p. 558.
6. UNEP, "Revised Draft Protocol on Chlorofluorocarbons Submitted by the U.S.," UNEP/WG.151/L.2, Geneva, November 25, 1986.
7. Richard E. Benedick, press statement, Vienna, February 27, 1987; *New Scientist,* March 5, 1987, p. 17.
8. *Schutz der Erdatmosphäre,* pp. 209, 211.
9. Mostafa K. Tolba, "Nowhere to Hide," UNEP press release, Geneva, April 27, 1987.
10. "Through Rose-Colored Sunglasses," editorial, *New York Times,* May 31, 1987; for

the draft text, see UNEP, "Report of the Ad Hoc Working Group on the Work of Its Third Session, UNEP/WG.172/2, Geneva, May 8, 1987, pp. 15–17.

11. Thomas Netter, "31 Nations Agree to Protect Ozone," *New York Times*, May 1, 1987.
12. UNEP, "Seventh Revised Draft Protocol," UNEP/IG.79/3, July 15, 1987.
13. Mostafa K. Tolba, "Weighing the Cost of Compromise," UNEP press release, Montreal, September 14, 1987.
14. The anecdotes that follow are based on the author's personal involvement.
15. Mostafa K. Tolba, "Facing a Distant Threat," UNEP press release, Montreal, September 16, 1987.

7. *Points of Debate*

1. UNEP, "Report of the Ad Hoc Working Group on the Work of Its Third Session," UNEP/WG.172/2, Geneva, May 8, 1987, p. 15.
2. P. M. DuPuy, "The World Ceiling Production System of CFC and Its Advantages" (Paper submitted to UNEP Workshop on the Control of Chlorofluorocarbons, Leesburg, Va., September 8–12, 1986), p. 3. DuPuy's phrase echoes that of Alexis de Tocqueville, cited in Chapter 3, note 35.
3. UNEP, "Ad Hoc Scientific Meeting to Compare Model-Generated Assessments of Ozone Layer Change for Various Strategies for CFC Control," Würzburg, April 9–10, 1987, UNEP/WG.167/INF.1.
4. UNEP, "Seventh Revised Draft Protocol," UNEP/IG.79/3, July 15, 1987, p. 4.
5. See the discussion in Johan G. Lammers, "Efforts to Develop a Protocol on Chlorofluorocarbons to the Vienna Convention for the Protection of the Ozone Layer," *The Hague Yearbook of International Law* 1 (1988):244.
6. UNEP, "Revised Draft Protocol on Chlorofluorocarbons Submitted by the U.S.," UNEP/WG.151/L.2, Geneva, November 25, 1986, p. 1.
7. German Delegation to the Council of the European Communities, memorandum, "Verhandlungen für ein Protokoll über Fluorchlorkohlenwasserstoffe zum Wiener Übereinkommen zum Schutz der Ozonschicht," no. 4684/87, Brussels, February 12, 1987 (translation by author).
8. UNEP, "Ad Hoc Scientific Meeting."
9. UNEP, "Report of Ad Hoc Working Group," p. 7; and personal notes of the author.
10. UNEP, "Report of Ad Hoc Working Group," p. 17.
11. UNEP, "Seventh Revised Draft Protocol," p. 5.
12. Ibid., pp. 4–5.
13. Examples include voting in the World Bank, International Monetary Fund, International Maritime Organization, and International Wheat Agreement.
14. "U.S. Move to Weaken Plan on Ozone Is Seen," *New York Times*, September 9, 1987; Michael Weisskopf, "EPA Would Make Ratification of Ozone Pact More Difficult," *Washington Post*, September 9, 1987.
15. Johan G. Lammers, "Second Report of the International Committee on Legal Aspects of Long-Distance Air Pollution," International Law Association, Warsaw Conference, 1988, p. 12.

16. UNEP, "Proposal by the European Community," UNEP/WG.151/L.5, Vienna, February 17, 1987.
17. UNEP, "Report of Ad Hoc Working Group," p. 18.
18. Numerous sources for CFC production and use in various countries are cited in U.S. House of Representatives, Committee on Energy and Commerce, Hearings, March 9, 1987, *Ozone Layer Depletion* (Washington, D.C.: U.S. Government Printing Office, 1987), pp. 406–415.
19. John Temple Lang, "The Ozone Layer Convention: A New Solution to the Question of Community Participation in 'Mixed' International Agreements," *Common Market Law Review* 23 (1986):157.
20. U.S. State Department Report, 1985 (author's files).
21. Temple Lang, "The Ozone Layer Convention," pp. 160, 161.
22. Markus Jachtenfuchs, "The European Community and the Protection of the Ozone Layer," *Journal of Common Market Studies* 28, no. 3 (March 1990): 262.
23. Temple Lang, "The Ozone Layer Convention," p. 166.
24. Ibid., pp. 162, 173.
25. U.S. State Department report, 1985 (author's files).
26. *Message from the President of the United States Transmitting the Vienna Convention for the Protection of the Ozone Layer,* U.S. Senate Treaty Doc. 99-9 (Washington, D.C.: U.S. Government Printing Office, 1985), p. vi; Jachtenfuchs, "European Community and Protection," pp. 263–264.
27. Hal Collums, U.S. Department of State, Treaty Division, memorandum to Deborah D. Kennedy, attorney, U.S. Department of State, July 23, 1987; Deborah D. Kennedy, letter to author, December 31, 1987 (author's files).
28. Winfried Lang, "Diplomatie zwischen Ökonomie und Ökologie: Das Beispiel des Ozonvertrags von Montreal," *Europa Archiv: Zeitschrift für Internationale Politik,* February 2, 1988, p. 107 (translation by author); see also Environmental Data Services Ltd., *ENDS Report 152,* September 1987, pp. 23–24.

8. The Immediate Aftermath

1. See U.S. Congress, Office of Technology Assessment, "An Analysis of the Montreal Protocol on Substances That Deplete the Ozone Layer," Revised staff paper, Washington, D.C., 1988, pp. 31–41.
2. *Schutz der Erdatmosphäre: Eine internationale Herausforderung,* Zwischenbericht der Enquete-Kommission des 11. Deutschen Bundestages (Bonn, 1988), pp. 257–265.
3. A. B. Jaafar, "Trade War by Environmental Decree," *Asia Technology,* January 1990, p. 51.
4. U.S. Office of Technology Assessment, "Analysis of the Montreal Protocol," pp. 9–12, 37–38; J. T. B. Tripp, "The UNEP Montreal Protocol: Industrialized and Developing Countries Sharing the Responsibility for Protecting the Stratospheric Ozone Layer," *Journal of International Law and Politics* 20, no. 3 (1988):741.
5. Information obtained from private industry sources.
6. See "U.S. and Soviets Spur Scientific Collaboration," *Conservation Foundation Letter,* no. 1, 1988.

7. Eric J. Lerner, "International Cooperation in Space," *Aerospace America,* June 1989, p. 34.
8. "An Exemplary Ozone Agreement," *Newsweek,* September 28, 1987; "The Ozone Treaty," editorial, *Washington Post,* September 18, 1987. A sampling of editorial comment is collected in Alliance for Responsible CFC Policy, *The Montreal Protocol: A Briefing Book* (Rosslyn, Va., 1987), sec. III.
9. "Dozens of Nations Approve Accord to Protect Ozone," *New York Times,* September 16, 1987.
10. See, for example, "A Tentative Stab at Saving Ozone," editorial, *San Antonio Express News,* September 27, 1987; "Mending the Ozone Shield," editorial, *Cleveland Plain Dealer,* September 24, 1987.
11. Cheryl Sullivan, "Trend-setting Berkeley Takes Action on Behalf of Earth's Ozone Layer," *Christian Science Monitor,* October 7, 1987.
12. Alliance for Responsible CFC Policy, *The Montreal Protocol,* p. I-8.
13. Ibid., p. I-9.
14. Alan S. Miller, "Incentives for CFC Substitutes: Lessons for Other Greenhouse Gases," in *Coping with Climate Change,* ed. John C. Topping, Jr. (Washington, D.C.: Climate Institute, 1989), p. 549.
15. Sylvie Faucheux and J.-F. Noël, *Did the Ozone War End in Montreal?* Université de Paris, Centre Economie-Espace-Environnement, English digest (Paris: Cahiers du C.3.E., 1988), p. 11.
16. ISC Chemicals, "Chlorofluorocarbons and 'The Ozone Layer': The Facts and the Fiction," brochure, Bristol, U.K., 1987.
17. U.K. House of Lords, *Hansard* 500 (October 20, 1988): col. 1310.
18. Environmental Data Services Ltd., *ENDS Report 157,* February 1988, p. 5.
19. U.K. House of Lords, Session 1987–88, Select Committee on the European Communities, "Minutes of Evidence, 21 June 1988," in *The Ozone Layer: Implementing the Montreal Protocol* (London: Her Majesty's Stationery Office, 1988), pp. 2–3; *ENDS Report 157,* p. 5.
20. See, for example, Philip Shabecoff, "Industry Acts to Save Ozone," *New York Times,* March 21, 1988; idem, "Race for Substitutes to Help Save Ozone," ibid., March 31, 1988; Pamela S. Zurer, "Search Intensifies for Alternatives to Ozone-Depleting Halocarbons," *Chemical and Engineering News,* February 8, 1988.
21. "CFC Producers from Seven Nations Plan to Pool Knowledge, Jointly Conduct Tests," *International Environment Reporter* 11, no. 2 (February 1988): 110; "Allied-Signal, French Firm Agree to Jointly Develop Alternatives for CFCs," ibid., no. 4 (April 1988): 227–228; Cynthia P. Shea, *Protecting Life on Earth: Steps to Save the Ozone Layer,* Worldwatch Paper 87 (Washington, D.C.: Worldwatch Institute, 1988), p. 30.
22. EC Commission, "Meeting of National Experts: 7/8 December 1987, Implementation of the Montreal Protocol on Substances That Deplete the Ozone Layer," Working Document of the Commission Services, XI/III/997/87-EN, Brussels.
23. Netherlands Ministry of Housing, Physical Planning, and the Environment, "Memorandum on CFC Policy," The Hague, March 1988, p. 4.
24. Note by the United Kingdom, "Some Considerations regarding Community Ful-

fillment of Production Control Obligations," December 7, 1987, attached to EC Commission, "Meeting of National Experts."

25. Nigel Haigh, letter to the author, August 3, 1988. See also Haigh, *EEC Environmental Policy and Britain*, 2d ed. (London: Longman, 1989), p. 269; S. Clinton Davis, speech to the Ozone Depletion Conference, Royal Institute of British Architects, London, November 28, 1988, pp. 7–9; U.K. House of Lords, "Seventeenth Report, 12 July 1988," in *The Ozone Layer*, p. 7.
26. Calculated from Chemical Manufacturers Association, "Production, Sales, and Calculated Release of CFC-11 and CFC-12 through 1987," Washington, D.C., 1988, and from David W. Pearce, "The European Community Approach to the Control of Chlorofluorocarbons" (Paper submitted to United Nations Environment Programme Workshop on the Control of Chlorofluorocarbons, Leesburg, Va., September 8–12, 1986), p. 16.
27. U.K. House of Lords, "Seventeenth Report," p. 12.
28. Markus Jachtenfuchs, "The European Community and the Protection of the Ozone Layer," *Journal of Common Market Studies* 28, no.3 (March 1990): 275, 267.
29. Ibid., p. 268.
30. U.K. House of Lords, "Seventeenth Report," p. 12.
31. "Comments of Eurimpact (a CFC user group) to the Commission of the European Communities on Implementation of Montreal Protocol," Brussels, November 9, 1987, pp. 30, 37.

9. New Science, New Urgency

1. NASA, NOAA, National Science Foundation, and Chemical Manufacturers Association, "Initial Findings from Punta Arenas, Chile, Airborne Antarctic Ozone Experiment," fact sheet, September 30, 1987.
2. The author chaired this meeting.
3. Robert T. Watson, F. Sherwood Rowland, and John Gille, "Ozone Trends Panel Executive Summary," NASA, Washington, D.C., 1988; Richard A. Kerr, "Stratospheric Ozone Is Decreasing," *Science* 239 (1988): 1489–91.
4. "Ozone Trends Panel: Press Conference," NASA, Washington, D.C., March 15, 1988, p. 19; NASA, *Present State of Knowledge of the Upper Atmosphere 1988* (Washington, D.C., 1988).
5. F. Sherwood Rowland, "Chlorofluorocarbons and the Depletion of Stratospheric Ozone," *American Scientist* 77 (January–February 1989):42–44.
6. Richard E. Heckert, chief executive officer of E. I. du Pont de Nemours & Co., letter to Senators Robert Stafford, Max Baucus, and David Durenberger, March 4, 1987, cited in Forest Reinhardt, *Du Pont Freon Products Division*, case study prepared for Harvard Business School (Washington, D.C.: National Wildlife Federation, 1989), p. 15.
7. Philip Shabecoff, "Du Pont to Halt Chemicals That Peril Ozone," *New York Times*, March 25, 1988; William Glaberson, "Behind Du Pont's Shift on Loss of Ozone Layer," ibid., March 27, 1988; Cynthia P. Shea, "Why Du Pont Gave Up $600

Million," ibid., April 10, 1988; J. M. Steed, "Global Cooperation, Not Unilateral Action," *Environmental Forum*, July–August 1988, p. 15.

8. "Thomas Urges Tightening Montreal Protocol, Suggests Near-Total CFC Phaseout as Goal," *International Environment Reporter* 11, no. 9 (September 1988):465–466; Philip Shabecoff, "EPA Chief Asks Total Ban on Ozone-Harming Chemicals," *New York Times*, September 27, 1988.
9. UNEP, *The Ozone Layer* (Nairobi, 1987), p. 25.
10. David Doniger, "Politics of the Ozone Layer," *Issues in Science and Technology*, Spring 1988; John Gribbin, *The Hole in the Sky* (New York: Bantam, 1988), pp. 165, 171.
11. "Provisions of Protocol Too Little, Too Late for Protection of Ozone Layer, Rowland Says," *International Environment Reporter* 11, no. 4 (April 1988): 227; Friends of the Earth International, "UNEP: 'No Emergency,' *Atmosphere* 1, no. 2 (1988):1.
12. Richard E. Benedick, "A Double Threat to the Ozone Treaty," *International Herald Tribune*, June 19, 1987.
13. *Schutz der Erdatmosphäre: Eine internationale Herausforderung*, Zwischenbericht der Enquete-Kommission des 11. Deutschen Bundestages (Bonn, 1988).
14. Ibid., pp. 227–228.
15. Council of the European Communities, "Resolution for the Limitation of Use of Chlorofluorocarbons and Halons," Brussels, June 16, 1988, reprinted in U.K. House of Lords, Session 1987–88, Select Committee on the European Communities, "Seventeenth Report, 12 July 1988," in *The Ozone Layer: Implementing the Montreal Protocol* (London: Her Majesty's Stationery Office, 1988), app. 2, p. 16.
16. Environmental Data Services Ltd., *ENDS Report 161*, June 1988, pp. 5–6; United Kingdom Stratospheric Ozone Review Group, *Stratospheric Ozone 1988* (London: Her Majesty's Stationery Office, 1988).
17. *ENDS Report 161*, pp. 17–18; U.K. House of Lords, "Seventeenth Report," p. 12.
18. Virginia Bottomley, "Protecting the Ozone Layer: A Challenge for the World Community," speech to the Ozone Depletion Conference, Royal Institute of British Architects, London, November 28, 1988, p. 4.
19. Margaret Thatcher, speech to the Royal Society, September 27, 1988, released by London Press Service, Central Office of Information, VS075/88, September 28, 1988.
20. Bottomley, "Protecting the Ozone Layer," p. 9.
21. Environmental Data Services Ltd., *ENDS Report 169*, February 1989, p. 4; Markus Jachtenfuchs, "The European Community and the Protection of the Ozone Layer," *Journal of Common Market Studies* 28, no. 3 (March 1990):270.
22. Jachtenfuchs, "European Community and Protection," p. 268.
23. John Temple Lang, "The Ozone Layer Convention: A New Solution to the Question of Community Participation in 'Mixed' International Agreements," *Common Market Law Review* 23 (1986):161.
24. Benedick, "Double Threat to Ozone Treaty."
25. U.K. House of Lords, *Hansard* 500 (October 20, 1988): col. 1297.
26. L. J. Brinkhorst, "Ozone: Europe Is Moving," *International Herald Tribune*, June 21, 1988; Richard E. Benedick, "Europe and Ozone," ibid., June 30, 1988.
27. EC Commission, "Meeting of National Experts: 7/8 December 1987, Implementa-

tion of the Montreal Protocol on Substances That Deplete the Ozone Layer," Working Document of the Commission Services, XI/III/997/87-EN, Brussels, p. 3.

10. The Road to Helsinki

1. Imperial Chemical Industries, news release, Runcorn, Cheshire, November 22, 1988.
2. Kerry Knobelsdorff, "Industry Scrambles to Find Ozone-Safe CFC Substitutes," *Christian Science Monitor,* August 12, 1988; Laurie Hays, "Firms Intensify Race to Find Substitutes for Chemicals Linked to Ozone Depletion," *Wall Street Journal,* September 27, 1988; Environmental Data Services Ltd., *ENDS Report 169,* February 1989, pp. 4–6.
3. Pamela S. Zurer, "Search Intensifies for Alternatives to Ozone-Depleting Halocarbons," *Chemical and Engineering News,* February 8, 1988, p. 17; Laurie Hays, "CFC Curb to Save Ozone Will Be Costly," *Wall Street Journal,* March 28, 1988; Joel Kurtzman, "The Race to Commercialize Substitutes," *New York Times,* April 10, 1988.
4. Arjun Makhijani, Annie Makhijani, and Amanda Bickel, *Saving Our Skins: Technical Potential and Policies for the Elimination of Ozone-Depleting Chlorine Compounds* (Washington, D.C.: Environmental Policy Institute and Institute for Energy and Environmental Research, 1988), pp. 72–125; Cynthia P. Shea, *Protecting Life on Earth: Steps to Save the Ozone Layer,* Worldwatch Paper 87 (Washington, D.C.: Worldwatch Institute, 1988).
5. A. W. Trivelpiece et al., "Environmental, Health, and CFC-Substitution Aspects of the Ozone Depletion Issue," Oak Ridge National Laboratory, April 1989, p. 28.
6. Kerry Knobelsdorff, "Costs of Protecting Ozone: More-Expensive Appliances," *Christian Science Monitor,* April 1, 1988; Malcolm Browne, "In Protecting the Atmosphere, Choices Are Costly and Complex," *Los Angeles Times,* March 7, 1989; A. K. Naj, "Doubts Raised on Substitutes for CFCs; Problems Abound in Development, Use, as Ban Looms," *Wall Street Journal,* March 6, 1989; Richard Koenig, "Refrigeration Makers Plan for Future without CFCs," *Wall Street Journal,* December 15, 1989.
7. William Chandler, Howard Geller, and Marc Ledbetter, *Energy Efficiency: A New Agenda* (Washington, D.C.: American Council for an Energy-Efficient Economy, 1988), p. 24.
8. UNEP, "Scientific Assessment of Stratospheric Ozone: 1989, Executive Summary, Scientific Chapter Summaries, and Contributors," July 14, 1989, pp. 4, 11–14.
9. United Kingdom Stratospheric Ozone Review Group, *Stratospheric Ozone 1990* (London: Her Majesty's Stationery Office, 1990), p. 8; Michael J. Prather and Robert T. Watson, "Stratospheric Ozone Depletion and Future Levels of Atmospheric Chlorine and Bromine," *Nature* 344 (1990): 729.
10. Cicerone quoted in Philip Shabecoff, "Large Volcanic Eruption Could Damage Ozone, Two Researchers Report," *New York Times,* May 9, 1989; see also Susan Solomon and M.-R. Schoeberl, "Overview of the Polar Ozone Issue," *Geophysical Research Letters* 15 (1989):845–846.

11. UNEP, *Synthesis Report,* UNEP/OzL.Pro.WG II(1)/4, November 13, 1989, pp. 29–31; Stratospheric Ozone Review Group, *Stratospheric Ozone 1990,* pp. 11, 14–16.
12. UNEP, *Synthesis Report,* p. 28.
13. Ibid., p. 29.
14. UNEP, "Scientific Assessment of Stratospheric Ozone: 1989," pp. 25–31; UNEP, *Synthesis Report,* app. A.
15. "Safeguarding the Ozone Layer and the Global Climate from Chlorofluorocarbons and Related Compounds," statement issued by a group of environmental organizations, Helsinki, May 1, 1989.
16. "Saving the Ozone Layer: London Conference," Reference Services, Central Office of Information, London, no. 321/89, 1989; Nicholas Ridley, secretary of state for the environment, United Kingdom, memorandum to Kaj Bärlund, minister of environment, Finland, March 7, 1989 (printed report on the London Conference distributed to delegates at the First Meeting of Parties to the Montreal Protocol, Helsinki, May 1989).
17. J. C. Randal, "Third World Seeks Aid before Joining Ozone Pact," *Washington Post,* March 7, 1989; L. B. Stammer, "Saving the Earth: Who Sacrifices?" *Los Angeles Times,* March 13, 1989; "The Hole in the Ozone Logic," editorial, *South,* April 1989.
18. Mostafa K. Tolba, "The Need to Go Further: The Montreal Protocol 19 Months Later," UNEP press release, Helsinki, May 2, 1989.
19. UNEP, "Report of the Parties to the Montreal Protocol on the Work of Their First Meeting," UNEP/OzL.Pro.1/5, May 6, 1989, p. 20.
20. UNEP, Ad Hoc Working Group of Legal and Technical Experts (Working Group on Data Harmonization), "Report of the Second Session," The Hague, October 24–26, 1988, UNEP/OzL.WG.Data.2/3/Rev.2; UNEP, "Report of the Executive Director to the First Meeting of Parties to the Montreal Protocol on Substances That Deplete the Ozone Layer," UNEP/OzL.Pro.1/2, March 28, 1989, pp. 9–10.
21. For documentation of the preceding discussion, consult UNEP, "Report of Parties on Their First Meeting"; C. R. Whitney, "80 Nations Favor Ban to Help Ozone," *New York Times,* May 3, 1989.

11. The Protocol in Evolution

1. UNEP, *Synthesis Report,* UNEP/OzL.Pro.WG.II(1)/4, November 13, 1989, p. 28.
2. Ibid., pp. 6–9.
3. Environmental Data Services Ltd., *ENDS Report 177,* October 1989, p. 12. See also Joseph Scotto et al., "Biologically Effective Ultraviolet Radiation: Surface Measurements in the United States, 1974–85," *Science* 239 (1988): 762.
4. United Kingdom Stratospheric Ozone Review Group, *Stratospheric Ozone 1990* (London: Her Majesty's Stationery Office, 1990), p. 9.
5. UNEP, *Synthesis Report,* pp. 16–17.
6. UNEP, Open-Ended Working Group of the Parties to the Montreal Protocol, "Executive Director's Note," UNEP/OzL.Pro.Asmt.1/2, Nairobi, July 17, 1989, p. 8.
7. UNEP, *Synthesis Report,* pp. 9–12.
8. Ibid., p. 12.

9. Ibid., p. 13.
10. Environmental Data Services Ltd., *ENDS Report 177,* October 1989, pp. 15–16; *ENDS Report 181,* February 1990, pp. 7–8; John Holusha, "Ozone Issue: Economics of a Ban," *New York Times,* January 11, 1990; Malcolm Browne, "Grappling with the Cost of Saving Earth's Ozone," ibid., July 17, 1990.
11. Du Pont cited in Debora MacKenzie, "Cheaper Alternatives for CFCs," *New Scientist,* June 30, 1990, pp. 39–40; see also Alliance for Responsible CFC Policy, "Realistic Policies on HCFCs Needed in Order to Meet Global Ozone Protection Goals," brochure, Washington, D.C.: June 1990, p. 1.
12. Quoted in Browne, "Grappling with the Cost."
13. Imperial Chemical Industries (ICI), "The Development of Alternatives," in "The Ozone Issue and Regulation," brochure, Runcorn, Cheshire, June 1990.
14. Browne, "Grappling with the Cost."
15. MacKenzie, "Cheaper Alternatives for CFCs."
16. "Managing Earth's Resources," special section, *Business Week,* June 18, 1990, pp. 36, 40.
17. John Maggs, "U.S. Has Jump Start on CFC Ban," *Journal of Commerce,* June 25, 1990.
18. Alliance for Responsible CFC Policy, "Realistic Policies on HCFCs," p. 3.
19. UNEP, *Synthesis Report,* pp. 35, 47, 48.
20. Ibid., p. 35.
21. ICI, "HCFCs—The Low ODP Solution," in "The Ozone Issue and Regulation."
22. Alliance for Responsible CFC Policy, "Realistic Policies on HCFCs," pp. 1–6.
23. Quoted in Friends of the Earth, "Briefing Sheet: Methyl Chloroform—Ozone Destroyer," London, June 1990.
24. *ENDS Report 177,* pp. 13–16; *ENDS Report 181,* pp. 7–8.
25. UNEP, *Synthesis Report,* p. 37.
26. *ENDS Report 177,* p. 14.
27. Natural Resources Defense Council, *Wanted: For Destruction of the Ozone Layer* (Washington, D.C., 1990), p. 7.
28. *ENDS Report 181,* p. 7.
29. UNEP, Second Meeting of the Parties to the Montreal Protocol, "List of Meetings and Documents," UNEP/OzL.Pro.2/Inf.2, London, June 7, 1990.
30. Vienna Convention, art. 9, para. 1; Montreal Protocol, art. 2, paras. 9 and 10. See Appendixes A and B.
31. UNEP, Open-Ended Working Group of the Parties to the Montreal Protocol, "Report of the Legal Drafting Group," UNEP/OzL.Pro.WG.II(1)/5, Geneva, November 20, 1989, p. 2.
32. UNEP, Open-Ended Working Group of the Parties to the Montreal Protocol, Second Session of the First Meeting, "Final Report," UNEP/OzL.Pro.WG.I(2)/4, Nairobi, September 4, 1989, p. 2.
33. UNEP, "Report of the First Session of the Bureau of the Parties to the Montreal Protocol," UNEP/OzL.Pro.Bur.1/2, Geneva, September 29, 1989.
34. UNEP, Open-Ended Working Group of the Parties to the Montreal Protocol, "Report of the Mechanical Working Group," UNEP/OzL.Pro.WG.II(1)/6, Geneva, November 22, 1989, p. 2; UNEP, "Report of the Legal Drafting Group," pp. 20–25.

35. See Montreal Protocol, art. 2, para. 10; and Vienna Convention, art. 9.
36. For further discussion of these potential difficulties, see UNEP, "Report of Mechanical Working Group," paras. 1, 2, 14, 19, 25, and 26.

12. The South Claims a Role

1. J. T. B. Tripp, "The UNEP Montreal Protocol: Industrialized and Developing Countries Sharing the Responsibility for Protecting the Ozone Layer," *Journal of International Law and Politics* 20, no. 3 (1988):744.
2. "The Hole in the Ozone Logic," editorial, *South,* April 1989; J. C. Randal, "Third World Seeks Aid before Joining Ozone Pact," *Washington Post,* March 7, 1989.
3. UNEP, *Synthesis Report,* UNEP/OzL.Pro.WG.II(1)/4, November 13, 1989, p. 14.
4. Ibid.
5. U.S. Congress, Office of Technology Assessment, "An Analysis of the Montreal Protocol on Substances That Deplete the Ozone Layer," Revised staff paper, Washington, D.C., 1988, pp. 9, 12.
6. UNEP, Open-Ended Working Group of the Parties to the Montreal Protocol, Second Session of the First Meeting, "Final Report," UNEP/OzL.Pro.WG.I(2)/4, Nairobi, September 4, 1989, p. 11. See also Tripp, "The UNEP Montreal Protocol," p. 741.
7. James L. Tyson, "Why China Says Ozone Must Take Back Seat in Drive to Prosperity," *Christian Science Monitor,* March 23, 1989.
8. UNEP, *Synthesis Report,* p. 10.
9. Ibid., pp. 34, 36. See also Irving M. Mintzer, William Moomaw, and Alan S. Miller, *Saving the Shield: Strategies for Phasing Out Chlorofluorocarbons* (Washington, D.C.: World Resources Institute, 1989), p. 2.
10. D. F. Kohler, John Haaga, and Frank Camm, *Projections of Consumption of Products Using Chlorofluorocarbons in Developing Countries* (Santa Monica, Calif.: Rand Corporation, 1987).
11. UNEP, *Synthesis Report,* pp. 8, 15.
12. UNEP, Open-Ended Working Group of the Parties to the Montreal Protocol, First Session of the First Meeting, "Final Report," UNEP/OzL.Pro.WG.I(1)/3, Nairobi, August 25, 1989, pp. 4, 6, 8, 9.
13. UNEP, Open-Ended Working Group of the Parties to the Montreal Protocol, "Report of the Legal Drafting Group," UNEP/OzL.Pro.WG.II(1)/5, Geneva, November 20, 1989, pp. 14, 16, 17.
14. UNEP, Open-Ended Working Group of the Parties to the Montreal Protocol, "Report of the Second Session of the Second Meeting," UNEP/OzL.Pro.WG.II(2)/7, Geneva, March 5, 1990.
15. Ibid., p. 12.
16. Ibid., p. 3.
17. Kevin Fay, "Statement before the House Committee on Science, Space and Technology," Alliance for Responsible CFC Policy, Washington, D.C., July 11, 1990, p. 6; International Chamber of Commerce, International Environmental Bureau, *Newsletter,* no. 23, Geneva, May 1990.

18. UNEP, Industry and Environment Office, "Informal Consultative Meeting with Industry, May 8, 1990, Draft Summary Report," Paris, May 1990, pp. 4, 6–8.
19. UNEP, Open-Ended Working Group of the Parties to the Montreal Protocol, "Letter of 30 April 1990 from Barber B. Conable, President of the World Bank, to Dr. M. K. Tolba, Executive Director of UNEP," UNEP/OzL.Pro.WG.III(2)/Inf.8, Geneva, May 7, 1990.
20. Michael Weisskopf, "U.S. Intends to Oppose Ozone Plan," *Washington Post,* May 9, 1990.
21. UNEP, "Final Report," August 25, 1989, pp. 2, 4.
22. Philip Shabecoff, "U.S. Is Assailed at Geneva Talks for Backing Out of Ozone Plan," *New York Times,* May 10, 1990; Richard Darman, *Keeping America First: American Romanticism and the Global Economy,* Second Annual Albert H. Gordon Lecture, Harvard University (Washington, D.C.: Office of Management and Budget, 1990), p. 4.
23. "A Baffling Ozone Policy," *Time,* May 21, 1990; "A Serious Mistake on CFCs," editorial, *Washington Post,* May 11, 1990.
24. Shabecoff, "U.S. Is Assailed"; Michael Weisskopf, "U.S. Drops Opposition to CFC Phaseout Fund," *Washington Post,* June 16, 1990.
25. Letter to President George Bush from Senators John H. Chafee, Slade Gorton, William S. Cohen, John Heinz, John C. Danforth, Robert W. Kasten, Jr., John W. Warner, Rudy Boschwitz, Warren B. Rudman, Dave Durenberger, Pete V. Domenici, and James M. Jeffords, May 9, 1990; letter to President George Bush from Senators Al Gore, Alan Cranston, Patrick J. Leahy, John F. Kerry, Joseph Lieberman, Max Baucus, Claiborne Pell, Joseph R. Biden, Jr., Frank R. Lautenberg, Timothy E. Wirth, Brock Adams, Bill Bradley, May 9, 1990. *Congressional Record: Senate,* May 22, 1990, pp. S6759–60, and author's files.
26. *Congressional Record: Senate,* May 22, 1990, p. S6759.
27. UNEP, Open-Ended Working Group of the Parties to the Montreal Protocol, "Report of the Second Session of the Third Meeting," UNEP/OzL.Pro.WG.III(2)/3, Geneva, May 22, 1990, p. 4.
28. UNEP, "Report of the Third Meeting of the Bureau of the Montreal Protocol," UNEP/OzL.Pro.Bur.3/2, Geneva, May 15, 1990, p. 3.

13. Strong Decisions in London

1. UNEP, Second Meeting of the Parties to the Montreal Protocol, "Proposed Adjustments and Amendments to the Control Measures of the Montreal Protocol—Revised Note by the Executive Director," UNEP/OzL.Pro.WG.IV/2/Rev./1, London, June 20, 1990.
2. Council of the European Communities, "Meeting Document," CONS/ENV/90/7, Luxembourg, June 7, 1990.
3. Der Bundesminister für Umwelt, Naturschutz und Reaktorsicherheit, "Pressemitteilung," 77/90, May 30, 1990; Federal Republic of Germany, Federal Environment Agency, *Responsibility Means Doing Without—How to Rescue the Ozone-Layer* (Berlin, 1989).
4. John Holusha, "Du Pont to Construct Plants for Ozone-Safe Refrigerant," *New*

York Times, June 23, 1990; Imperial Chemical Industries (ICI), "The Ozone Issue and Regulation," brochure, Runcorn, Cheshire, June 1990.

5. See, for example, Friends of the Earth, *Funding Change: Developing Countries and the Montreal Protocol,* pamphlet (n.p., 1990; Greenpeace International, *The Failure of the Montreal Protocol 1990,* pamphlet (n.p., 1990); Natural Resources Defense Council (NRDC), "The New Montreal Protocol—Will It Close the Holes in the Ozone Treaty?" Washington, D.C., 1990.
6. UNEP, *Synthesis Report,* UNEP/OzL.Pro.WG.II(1)/4, Geneva, November 13, 1989, p. 11.
7. ICI, *Environmental Issues* (ICI Chemicals and Polymers: Runcorn, Cheshire, 1990). For examples involving methyl chloroform, see Chapter 11.
8. Liz Cook, "Testimony before the Natural Resources, Agricultural Research, and Environment, and International Scientific Subcommittees of the Committee on Science, Space and Technology, U.S. House of Representatives," Friends of the Earth, Washington, D.C., July 11, 1990, p. 9.
9. ICI, "The Ozone Issue and Regulation."
10. Article 9 of the Vienna Convention provides that protocol amendments will enter into force following ratification by two-thirds of the parties, unless the parties agree otherwise. Because of the growing number of parties, it was determined that waiting for two-thirds to ratify would unnecessarily delay the process. Hence, the figure of 20 parties was chosen at London, and amendment of the Vienna Convention was also recommended.
11. Statement by the Chief of Staff, Washington, The White House, June 15, 1990.
12. Michael Weisskopf, "U.S. Drops Opposition to CFC Phaseout Fund," *Washington Post,* June 16, 1990; Philip Shabecoff, "U.S. to Back Fund to Protect Ozone," *New York Times,* June 16, 1990; C. R. Whitney, "Poll Tax in This Aide's Environment," ibid., June 19, 1990.
13. UNEP, "Protecting the Ozone Layer—A Resounding Success," press release, London, July 1990.
14. UNEP,"Report of the Second Meeting of the Parties to the Montreal Protocol on Substances That Deplete the Ozone Layer," UNEP/OzL.Pro.2/3, London, June 29, 1990, p. 7.
15. Michael J. Prather and Robert T. Watson, "Stratospheric Ozone Depletion and Future Levels of Atmospheric Chlorine and Bromine," *Nature* 344 (1990): 732.
16. UNEP, *Synthesis Report,* pp. 10–11.
17. Jim Fuller, "U.S. Urges Control of Two Ozone-Damaging Chemicals," *U.S. Information Agency Wire Service,* June 29, 1990.
18. UNEP, *Synthesis Report,* pp. 37, 10–11.
19. Cited in NRDC, "The New Montreal Protocol," p. 4.
20. Cook, "Testimony," pp. 7–10.
21. Federal Republic of Germany, *Responsibility Means Doing Without,* pp. 60–61, 63.
22. Ibid., p. 62; Bundesminister für Umwelt, Naturschutz und Reaktorsicherheit, "Pressemitteilung," p. 6.
23. UNEP, Second Meeting of the Parties to the Montreal Protocol, "Report of the Executive Director," UNEP/OzL.Pro.2/2/Add.4/Rev.1, London, May 28, 1990, p. 4.

24. UNEP, Second Meeting of the Parties to the Montreal Protocol, "Report of the Executive Director," UNEP/OzL.Pro.2/2, London, March 30, 1990, p. 2.
25. UNEP, "Report of Executive Director," May 28, 1990, p. 5.
26. UNEP, Open-Ended Working Group of the Parties to the Montreal Protocol, "Remaining Issues to Be Addressed at the Fourth Meeting of the Working Group," UNEP/OzL.Pro.WG.IV/3, London, May 23, 1990, p. 2.
27. "Statement of the U.S. Delegation regarding the Financial Mechanism," London, June 20, 1990.
28. UNEP, "Revised Draft Report of the Fourth Meeting of the Open-Ended Working Group of the Parties to the Montreal Protocol," UNEP/OzL.Pro.WG.IV/L.1/Rev.1, London, June 28, 1990, p. 1.
29. Quoted in L. B. Stammer, "Ozone Victory Spurs War on Global Heating," *Los Angeles Times*, July 2, 1990.
30. UNEP, "Report of Second Meeting of Parties," p. 9.
31. L. B. Stammer, "Chinese Delegates to Seek Beijing's Approval for Pact to Protect Ozone," *Los Angeles Times*, June 29, 1990.
32. UNEP, Second Meeting of the Parties to the Montreal Protocol, "Draft Amendment to the Montreal Protocol on Substances That Deplete the Ozone Layer," UNEP/OzL.Pro.2/L.4/Rev.1, London, June 29, 1990, p. 9.
33. Stammer, "Chinese Delegates."
34. UNEP, "Protecting the Ozone Layer."
35. UNEP, "Report of Second Meeting of Parties," pp. 2, 9.
36. Kevin Fay, "Statement before the House Committee on Science, Space and Technology," Alliance for Responsible CFC Policy, Washington, D.C., July 11, 1990, p. 9.
37. Quoted in Stammer, "Ozone Victory."

14. Looking Ahead: A New Global Diplomacy

1. Robert T. Watson, M. A. Geller, Richard S. Stolarski, and R. F. Hampson, *Present State of Knowledge of the Upper Atmosphere* (Washington, D.C.: National Aeronautics and Space Administration, 1986), p. 18.
2. Roger T. Revelle and Hans E. Suess, "Carbon Dioxide Exchange between Atmosphere and Ocean and the Question of an Increase of Atmospheric CO_2 during the Past Decades," *Tellus* 9 (1957):19.
3. See UNEP, *The Greenhouse Gases* (Nairobi, 1987); and International Council of Scientific Unions, UNEP, and World Meteorological Organization, *Report of the International Conference on the Assessment of the Role of Carbon Dioxide and of Other Greenhouse Gases in Climate Variations and Associated Impacts,* Villach, Austria, October 9–15, 1985 (Geneva: WMO, 1986).
4. World Resources Institute, *World Resources 1987* (Washington, D.C., 1987), p. 178.
5. Robert Repetto, *Wasting Assets: Natural Resources in the National Income Accounts* (Washington, D.C.: World Resources Institute, 1989); Yusuf Ahmad, Salah El Serafy, and Ernst Lutz, eds., *Environmental Accounting for Sustainable Development* (Washington, D.C.: World Bank, 1989); Herman E. Daly and John B. Cobb, Jr., *For the Common Good* (Boston: Beacon Press, 1989), pt. 1.

6. The four episodes were: the abandonment of U.S. support for international controls by EPA Administrator Anne Gorsuch in 1981, reversed two years later during the negotiations for the Vienna Convention (see Chapter 4); the unsuccessful eleventh-hour attempt in 1985 to prevent the United States from signing the Vienna Convention (Chapter 4); the prolonged campaign in the first half of 1987 by certain officials to overturn the U.S. negotiating position on the Montreal Protocol, finally overruled by President Reagan (Chapter 5); and the decision on the U.S. position on the new ozone fund in May 1990, revised a month later just before the Second Meeting of Parties (Chapters 12 and 13).
7. Glenn Frankel, "Governments Agree on Ozone Fund," *Washington Post,* June 30, 1990.
8. "Draft Report Attributes One-Quarter of Greenhouse Effect to CFC Emissions," in *World Climate Change Report* (Washington, D.C.: Bureau of National Affairs, 1989), p. 15.
9. Calculated from World Resources Institute, *World Resources 1990–91* (New York: Oxford University Press, 1990), pp. 348–349.
10. Mostafa K. Tolba, "The Ozone Agreement—and Beyond," *Environmental Conservation* 14, no. 4 (1987):290.

Index

Institute for the Study of Diplomacy

Georgetown University School of Foreign Service

The Institute for the Study of Diplomacy concentrates on the *processes* of conducting foreign relations abroad, in the belief that studies of diplomatic operations are useful means of teaching or improving diplomatic skills and of broadening public understanding of diplomacy. Working closely with the academic program of the Georgetown University School of Foreign Service, the Institute conducts a program of research, publication, teaching, diplomats in residence, conferences, and lectures.

World Wildlife Fund & The Conservation Foundation

Board of Directors

World Wildlife Fund & The Conservation Foundation (WWF) is the largest private U.S. organization working worldwide to conserve nature. WWF works to preserve the diversity and abundance of life on Earth and the health of ecological systems by protecting natural areas and wildlife populations, promoting sustainable use of natural resources, and promoting more efficient resource and energy use and the maximum reduction of pollution. WWF is affiliated with the international WWF network, which has national organizations, associates, or representatives in nearly forty countries. In the United States, WWF has more than one million members.